Ahmed Memdouh YOUNSI

Physical,Digital,and Embedded Analysis of a Microcontroller-Based ECG

Ahmed Memdouh YOUNSI

Physical,Digital,and Embedded Analysis of a Microcontroller-Based ECG

Noor Publishing

Imprint

Cover image: www.ingimage.com

Publisher:
Noor Publishing
is a trademark of
Dodo Books Indian Ocean Ltd. and OmniScriptum S.R.L publishing group

120 High Road, East Finchley, London, N2 9ED, United Kingdom
Str. Armeneasca 28/1, office 1, Chisinau MD-2012, Republic of Moldova, Europe
Managing Directors: Ieva Konstantinova, Victoria Ursu
info@omniscriptum.com

Printed at: see last page
ISBN: 978-620-9-31321-9

ABSTRACT:

This work presents the design and implementation of a microcontroller-based electrocardiogram (ECG) acquisition system that integrates biomedical principles with embedded electronics to address the growing need for portable, low-cost, and real-time cardiac monitoring solutions, particularly in environments with limited access to advanced diagnostic infrastructure. The ECG signal, originating from the bioelectrical activity of myocardial cells, is captured using electrodes, amplified by a high-CMRR instrumentation amplifier, then isolated via optocouplers and filtered with bandpass filters to remove noise while preserving diagnostic features. The conditioned signal is digitized by the PIC16F877 microcontroller's 10-bit ADC module, enabling precise sampling and conversion of cardiac waveforms, and transmitted via the USART interface, using the MAX232 driver, to a PC for visualization and analysis via a custom interface developed in Visual Basic .NET. This multidisciplinary system demonstrates how to design and build functional biomedical devices. It lays the groundwork for future enhancements, including wireless transmission, cloud-based storage, and AI-assisted diagnostics. By bridging theoretical knowledge with practical implementation, this system promotes accessible and advanced healthcare technologies. This modest project focuses on studying and implementing an electrocardiogram (ECG), which records the electrical activity associated with cardiac contractions. The primary purpose of this device is to capture the heart's electrical signals.

Key words:

Electrocardiogram (ECG); PIC16F877; analog-to-digital conversion; embedded systems; cardiac monitoring.

摘要：

本研究介绍了一种基于微控制器的心电图（ECG）采集系统的设计与实现。该系统结合了生物医学原理与嵌入式电子技术，旨在应对对便携、低成本、实时心脏监测方案日益增长的需求，特别适用于诊断基础设施有限的环境。源自心肌细胞生物电活动的心电信号通过电极采集，经高共模抑制比（CMRR）的仪表放大器放大后，再通过光耦隔离，并借助带通滤波器滤除噪声，同时保留诊断特征。经过调理的信号由 PIC16F877 微控制器的 10 位 ADC 模块数字化，实现心脏波形的精确采样与转换，并通过 USART 接口，借助 MAX232 驱动器，传送至个人计算机，利用 Visual Basic .NET 开发的专用界面进行可视化与分析。该跨学科系统展示了如何设计和构建功能性生物医学设备，并为未来的进一步改进奠定了基础，包括无线传输、基于云的存储以及人工智能辅助诊断。通过将理论知识与实际应用相结合，系统推动了普惠与先进医疗技术的发展。本项目以研制和实现心电图（ECG）为核心，记录与心脏收缩相关的电活动。该设备的主要目标是捕捉心脏的电信号。

关键词：

心电图（ECG）；PIC16F877；模数转换；嵌入式系统；心脏监测。

Table of Contents

List of Figures:

List of tables :

General Introduction

General Introduction:

In recent decades, the field of medical electronics has undergone a profound transformation, becoming an indispensable pillar of modern healthcare[**1**]. The integration of electronic systems into medical diagnostics and therapeutic procedures has revolutionized the way clinicians assess, monitor, and treat patients[**2**]. From imaging technologies to implantable devices, electronics now permeate every level of clinical practice, offering precision, speed, and reliability that far surpass traditional methods[**3**].

Historically, medical diagnosis relied heavily on the physician's sensory perception and clinical intuition. The diagnostic process was primarily based on patient interviews, physical examination, and the interpretation of subjective symptoms. While these methods remain foundational, they are inherently limited by human perception and variability. The advent of electronic instrumentation has enabled the acquisition of objective, quantifiable physiological data, thereby enhancing diagnostic accuracy and enabling early detection of pathological conditions[**4**].

Among the earliest and most impactful applications of medical electronics is the **electrocardiogram (ECG)**—a non-invasive technique that records the electrical activity of the heart [**5**]. The ECG has become a cornerstone in cardiology, allowing clinicians to detect arrhythmias, myocardial infarctions, conduction abnormalities, and other cardiac pathologies with remarkable efficiency [**6**]. Its widespread use in hospitals, clinics, and even portable monitoring systems underscores its clinical value and technological maturity [**7**].

The ECG functions by capturing the bioelectrical signals generated by the depolarization and repolarization of myocardial fibers. These signals, though minute in amplitude, carry rich information about the physiological and pathological state of the heart. To accurately acquire and interpret these signals, a comprehensive understanding of both **cardiac electrophysiology** and **signal processing techniques** is essential [**8**]. This dual perspective—biomedical and engineering—forms the conceptual backbone of this work.

The present study is situated at the intersection of **biomedical science** and **embedded systems engineering**, aiming to design and implement a microcontroller-based ECG acquisition system. The project is motivated by the growing need for **portable, cost-effective, and real-time monitoring solutions**, particularly in resource-limited settings where access to advanced medical infrastructure may be constrained [**9**]. By leveraging the capabilities of the **PIC16F877 microcontroller**, this work seeks to demonstrate how embedded platforms can be harnessed to build functional and efficient biomedical devices [**10**].

The structure of this work reflects a progressive and interdisciplinary approach:

Chapter 1 presents a unified theoretical foundation by combining two essential domains: the **electrophysiological mechanisms of the heart** and the **principles of electrocardiography**. It begins with a review of cardiac anatomy and the conduction system, followed by a detailed analysis of action potentials, ion exchanges, and the generation of ECG waveforms. The chapter also explores the historical evolution of electrocardiography, the clinical significance of ECG components, and the standard lead configurations used in practice. Finally, it introduces the electronic and computational aspects of ECG acquisition, including amplification, filtering, and signal digitization.

Chapter 2 is dedicated to the **PIC16F877 microcontroller**, which serves as the core of the acquisition system. The chapter outlines the microcontroller's architecture, functional blocks, and programming environment, emphasizing its suitability for biomedical applications. Special attention is given to its analog-to-digital conversion capabilities, memory structure, and communication interfaces.

Chapter 3 delves into the **conversion and communication subsystems**. It examines the analog-to-digital conversion (ADC) process, which is critical for converting the analog ECG signal to a digital format suitable for processing and storage. The chapter also discusses serial communication protocols, particularly UART, for transmitting digitized signals to external devices such as personal computers or display modules.

Chapter 4 presents the **practical implementation** of the ECG acquisition system. It details the hardware integration, software development, and signal visualization process. The chapter includes experimental results, signal traces, and a discussion of challenges encountered during development, such as noise reduction, signal stability, and real-time performance. The system's performance is evaluated based on accuracy, responsiveness, and usability.

Through this multidisciplinary exploration, the present work aims to contribute to the growing body of knowledge in **biomedical instrumentation**, particularly in the context of low-cost, embedded ECG systems. It demonstrates how a solid understanding of physiological principles, combined with engineering innovation, can lead to the development of practical tools that enhance patient care and expand access to diagnostic technologies.

Ultimately, this project reflects a broader vision: to bridge the gap between **theoretical knowledge and real-world application**, and to empower future biomedical engineers and clinicians with the tools and insights needed to design the next generation of intelligent, accessible, and patient-centered medical devices.

Medical electronics is an indispensable tool for diagnosing and treating many diseases, playing a crucial role in clinical practice. Previously, diagnosis and medical monitoring relied on the results of a doctor-patient interview about their medical history and subjective symptoms, as well as the doctor's visual, auditory, and tactile examination of the patient.

Chapter 1

Electrophysiological Foundations and Theoretical Analysis of the Electrocardiogram (ECG)

I.1 Introduction:

The electrocardiogram (ECG) remains one of the most fundamental and widely used tools in clinical cardiology and biomedical engineering for assessing the electrical activity of the heart. A comprehensive understanding of ECG signals requires a dual perspective: the physiological mechanisms underlying cardiac electrogenesis, and the theoretical principles governing signal acquisition, interpretation, and processing. This chapter provides an integrated overview of cardiac electrophysiology and the theoretical foundations of electrocardiography [11]. It begins with a review of the anatomical and functional organization of the heart, emphasizing the specialized conduction system and the biophysical basis of action potentials. The chapter then transitions to the principles of ECG signal formation, including the historical evolution of electrocardiography, the clinical significance of ECG waveforms, and the standardized lead configurations used in practice.
Furthermore, the chapter explores the instrumentation and electronic components used in ECG acquisition, including electrodes, amplifiers, and analog-to-digital converters [12, 13]. Special attention is given to signal conditioning, noise filtering, and the computational techniques employed in modern ECG processing. This foundational knowledge sets the stage for the practical implementation of an embedded ECG acquisition system using a microcontroller, as detailed in the subsequent chapters.

I.2 Anatomy and function of the heart:

I.2.1 Presentation :

The heart is a hollow, muscular organ that receives blood from the veins and pumps it into the arteries, thereby ensuring circulation [14]. The human heart is approximately the size of a closed fist and weighs 250g in adults. It has a vaguely conical shape, with the base pointing upwards and to the right, slightly inclined backwards; the apex touches the chest wall between the fifth and sixth ribs.

The human heart comprises two parallel systems, each consisting of an atrium and a ventricle. Due to their anatomical positions, these systems are often referred to as the right and left heart [15, 16].

I.2.2 Electrical conduction of the heart:

This is an "electrical" network composed of cardiac cells that exhibit differentiated properties of conductivity and excitability [17]. The heart has two special types of muscle tissue: the myocardium and the nodal tissue. The nodal tissue includes the sinoatrial node, the anterior descending (Achoff-Tawara) node, the bundle of His, and the Purkinje fibers. The myocardium covers the entire remaining portion of the heart [18, 29]. Cardiac electrical activity originates in the sinoatrial node, located in the walls of the right atrium (figure I.1).

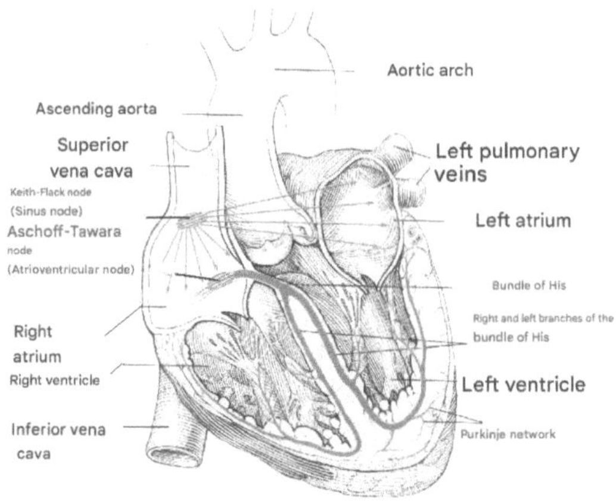

Figure I.1: The Heart's Electrical Conduction System.

I.3 Cardiac electrophysiology :

I.3.1 Myocardial fiber, activation, and electrochemistry:

There are two types of fibers, with different histologies :

-- Muscle fibers: contractile

-- Nodal tissue: conductive

The electrophysiological properties of the myocardial fiber, such as excitability, automaticity, and conduction, depend on interactions among the multiple electrical charges within the cellular environment. Figure I.2 represents the Polarization of a cardiac cell. When an electrical stimulus excites a cardiac cell, ions enter it through sodium, calcium, and potassium channels. At rest, the external surface of a cell is positively charged[30, 31].

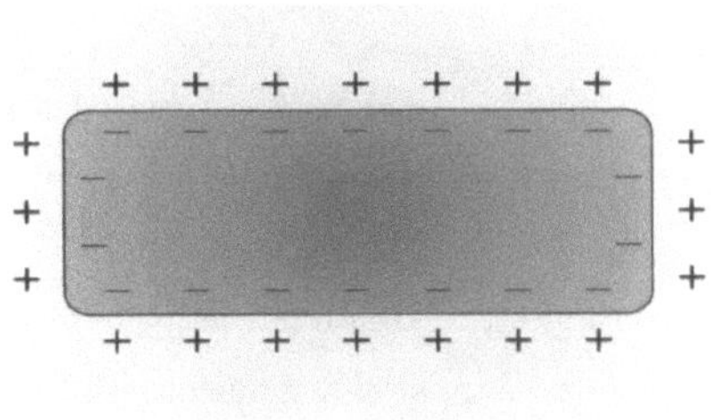

Figure I.2: Polarization of a cardiac cell.

During depolarization (figure I.3), sodium ions cross the membrane, and the outer surface then becomes negatively charged.

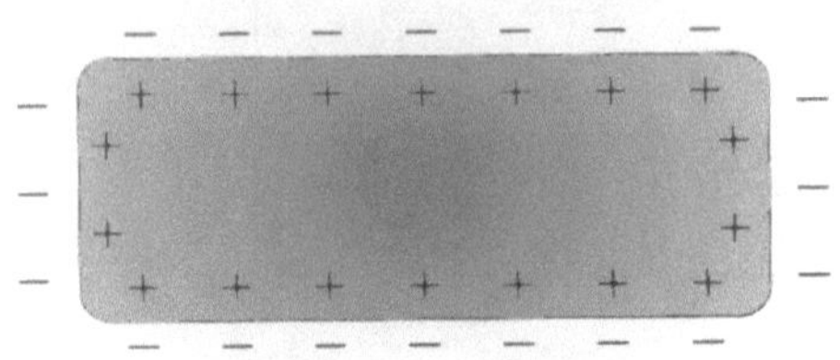

Figure I.3: Depolarization of a cardiac cell.

This depolarization propagates along the membrane, leading to the formation of multiple action potentials that diffuse and cause a reversal of the membrane potential. Figure I.4 shows that the high conductivity of the Na+ ion allows K+ to be expelled from the cell. The membrane thus regains the positivity of its external surface: this is repolarization. At rest, the intracellular composition of K+ is thirty times that of Na+.

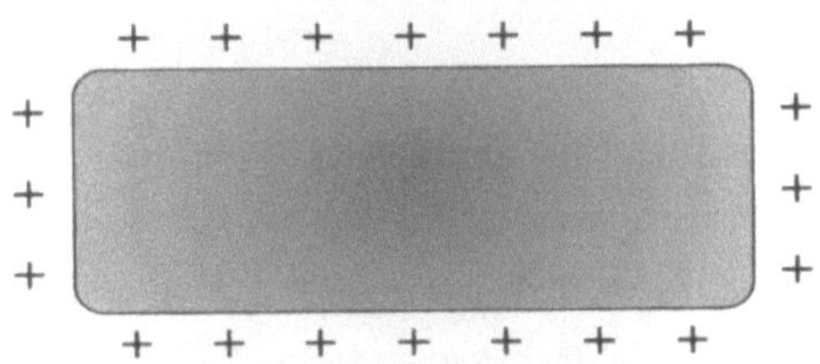

Figure I.4: Repolarization of a cardiac cell .

These exchanges, linked to the action potential, are passive and a result of transmembrane ion concentration gradients driven by the membrane's selective permeability; they do not require energy [32, 33].

In figure I.5, there are active exchanges that consume energy, allowing the resting potential to be restored by the resting transmembrane concentration gradients. This gives us a resting potential of -80 to -90 millivolts.

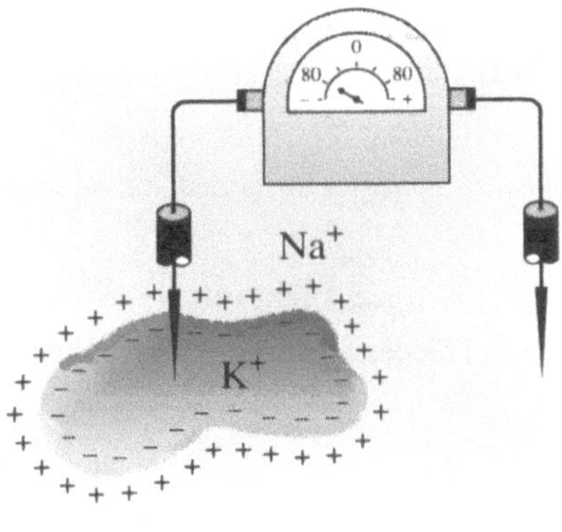

Figure I.5: Transmembrane resting potential of a cardiac cell.

I.3.2 Action potential and automatic center fibers :

Using a microelectrode, these depolarization and repolarization phenomena can be studied and graphically reproduced:

Figure I.6 indicates that the depolarization follows stimulation. It is the change in membrane potential from -90 mV (the resting potential, 0) to 0 mV in a few milliseconds. There is a threshold of membrane depolarization that must be reached for the cell to be fully depolarized. This threshold determines cellular excitability. The changes in membrane potential are significant enough for the impulse to propagate to neighboring cells and trigger further action potentials.

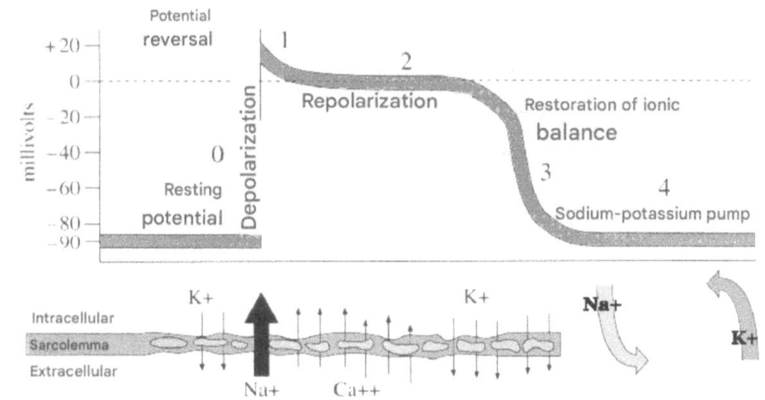

Figure I.6: the action potential phases and transmembrane ion movements.

Repolarization begins with a rapid initial repolarization phase (phase 1), driven by the inactivation of the incoming sodium current through fast sodium channels and by a weak chloride current. This is followed by a slow repolarization (phase 2), a slightly descending, oblique plateau linked to the cell's calcium current. Finally, a terminal repolarization phase (phase 3) occurs, characterized by a rapid drop in membrane potential corresponding to the outflow of potassium (this is the T wave of the surface ECG). During this entire repolarization period, the cell is no longer excitable: this is the refractory period.

Finally, between two depolarizations (phase 4), the cell has recovered its resting potential, with a positive charge on the surface, rich in sodium, compared to a charge poorer in potassium inside.

I.4 Electrocardiogram :

I.4.1 Generalities:

Electrocardiography records the electrical activity of the heart and produces electrocardiograms (ECGs). An electrocardiogram (ECG) is a graphic representation of the heart's electrical activity. This electrical activity is captured by electrodes placed on the skin. Each electrode detects the activation waves according to the plane of the heart it is scanning. The recording is made on graph paper.

The electrocardiogram is thus the recording on paper of the cardiac action currents transmitted to the body's surface. Two contact electrodes, connected by a wire to a galvanometer that measures current intensity, form a lead.

I.4.2 History of electrocardiography :

1842 The Italian physicist Carlo Matteucci shows that an electric current accompanies each contraction of the heart. Matteucci C. It is a physiological phenomenon produced by contracting muscles. Ann Chim Phys 1842; 6:339-341.In 1903, Einthoven negotiated commercial production of his device with Max Edelmann of Munich and with Horace Darwin of the Cambridge Scientific Instruments Company of London [24].

In 1942, Emanuel Goldberger added aVr, aVl, and aVf to Einthoven's frontal leads. This allowed him, along with the six precordial leads V1-V6, to create the first 12-lead electrocardiogram, which is still used today.

I.4.3 Medical use:

On an electrocardiographic tracing (figure I.7):

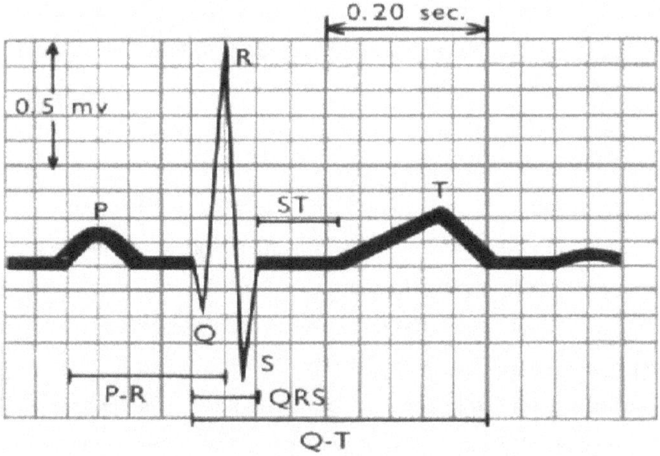

Figure I.7: tracing of an electrocardiogram.

P wave: It is the atrial depolarization wave. It is rounded in shape, often positive, of low amplitude (1 to 3 mV) and of less than 0.12 seconds.

QRS complex: It corresponds to the activation and depolarization of the ventricles preceding the mechanical effect of contraction; it consists of three segments:

Q wave: first negative deflection, **R wave:** first positive deflection, and **S wave:** negative deflection that follows the R wave.

The duration of the entire QRS complex varies from 0.06 to 0.1 seconds and is measured from the beginning of the QRS to the end of the S or R wave, as the case may be.

PR segment: This corresponds to the 1/10th of a second pause between atrial and ventricular activation, during the passage of the impulse from the atrioventricular node to the bundle of His. It is measured from the end of the P wave to the beginning of the QRS complex and corresponds to 0.03-0.04 seconds.

T wave: This is the ventricular repolarization period. It represents the inhibition of ventricular excitation. It is asymmetrical, with a slightly oblique ascending limb and a more abrupt descending limb. Its amplitude is less than 2 mm.

ST segment: This corresponds to the period of uniform ventricular excitation up to the ventricular recovery phase. It is measured from the end of the S or R wave to the beginning of the T wave. It usually is horizontal or slightly oblique and approximately isoelectric. An elevation or depression of more than 1 mm relative to the isoelectric line is abnormal.

U wave: This is evidence of a late repolarization of myocardial areas with an amplitude between that of the P wave and that of the T wave. It is less than 1/4 of the amplitude of the T wave.

PR interval: This is the atrioventricular conduction time. It is the time required for the impulse to depolarize the atria and then pass through the atrioventricular node and the trunk of the His bundle. It is measured from the onset of the P wave to the onset of the QRS complex. It is 0.12 to 0.23 seconds.

QT interval: This is the interval of depolarization (QRS), excitation (ST), and repolarization (T) of the ventricles. It is measured from the beginning of the QRS complex to the end of the T wave. The QT interval is a function of heart rate; therefore, it is preferable to use the corrected QT interval (QTc), which is calculated using Bazett's formula:

QTc = QT / Fc, which usually is less than or equal to 0.44 seconds.

I.4.4 Lead Systems:

The electrical potentials generated by the heart propagate throughout the body and appear on the body surface. The potential difference (PD) is measured at two points on the body surface using a pair of electrodes. Placing several pairs of electrodes in different positions yields different results because the heart's electrical field is spatially dependent. Measuring the cardiac vector in only one direction will not allow for its complete characterization.

Therefore, it is essential to use a standardized electrode placement (leads) for the clinical evaluation of the ECG signal. In practice, twelve leads are used in the frontal and transverse planes to explore the heart's electrical activity.

I.4.4.1 Frontal Leads :

These are "the limb leads": D1, D2, D3, aVR, aVL, and aVF

I.4.4.1.1 Bipolar Leads:

D1, D2 and D3 are bipolar derivations (figure I.8) which reflect the potential difference between two members.

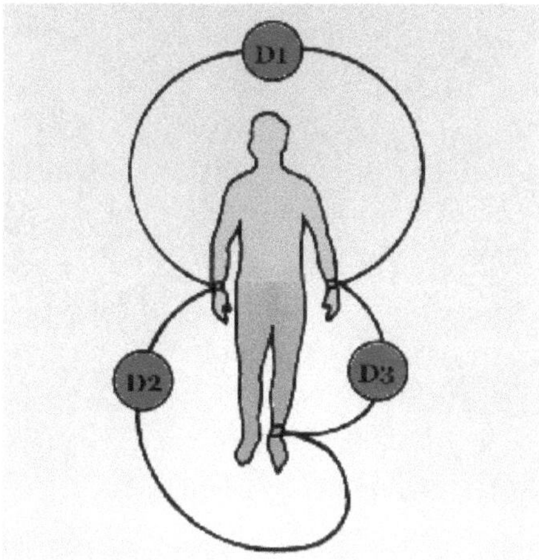

Figure I.8: Bipolar Leads D1, D2, D3.

D1, D2, and D3 describe the Einthoven triangle (figure I.9).

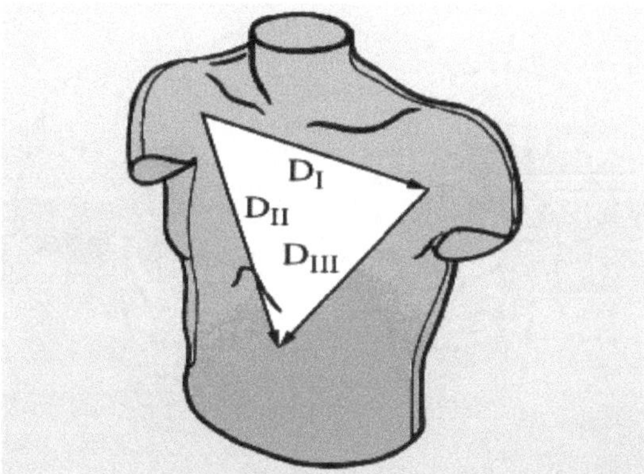

Figure I.9: Einthoven's triangle.

I.4.4.1.2 Unipolar leads:

aVR, aVL, and aVF are unipolar leads (figure I.10) and correspond to the limb with which they are connected, namely the right arm, left arm, and left leg, respectively.

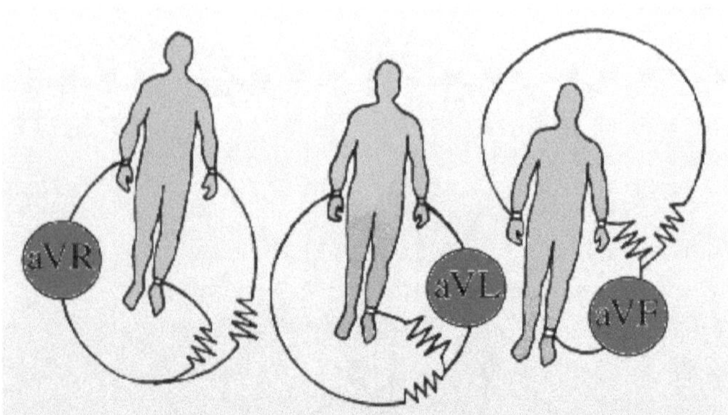

Figure I.10: Unipolar leads.

I.4.4.2 Precordial leads:

These are unipolar leads fixed at defined points on the chest wall (figure I.11).

For standard leads, they are named: V1 to V6:

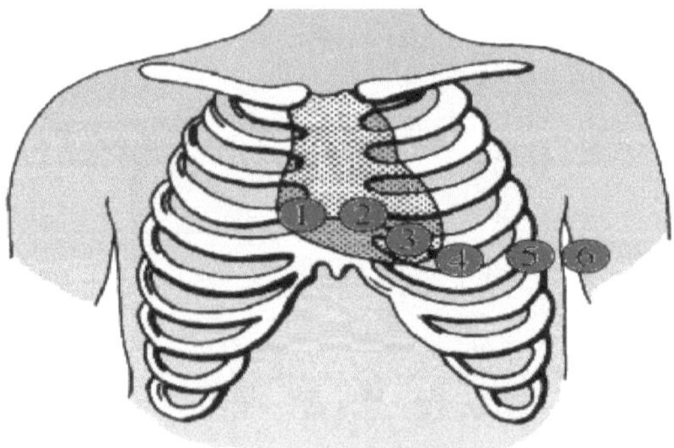

Figure I.11: Precordial leads.

I.5 Electrocardiograph :

I.5.1 Recording devices:

The recording device prints the tracing onto paper. Increasingly, the recording is digitized. The electrical current generated by the heart is conducted through the cables of the recording device, which has an amplifier for the transmitted signals and a galvanometer that moves a needle. The needle moves according to the magnitude of the electrical signal generated by the patient and records a positive, negative, or biphasic deflection [25, 31].

I.5.2 The electrodes:

The measuring electrodes are placed at various locations on the body, directly on the skin. The critical requirement for these electrodes is that they be relatively inert and non-polarizable. Silver plates coated with a layer of silver chloride (an insoluble gel) are good surface electrodes and the most commonly used (see figure I.12).

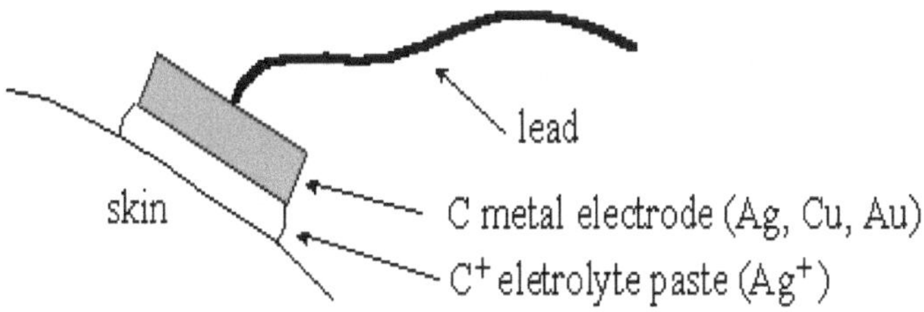

FigureI.12 : Bipotential electrodes.

Metal electrode (primarily silver), silver chloride gel. Electrodes used for electrocardiography in hospitals are of this type (figure I.13). These are disposable after a single use [20-23].

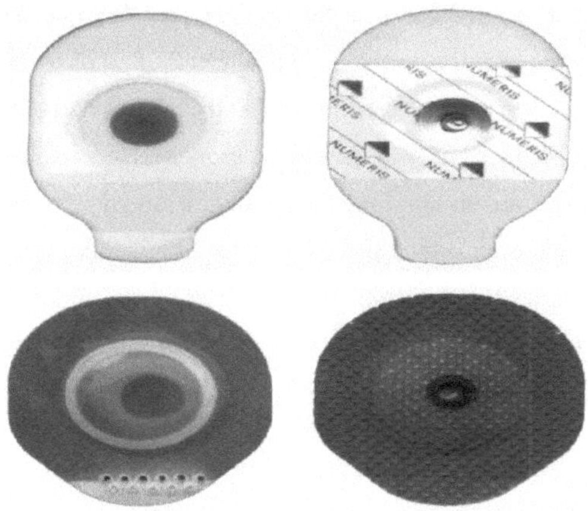

FigureI.13: Actual size approximately 5 cm maximum length.

I.5.3 Computer-based ECG acquisition chain:

The positioning of ECG acquisition electrodes has evolved along with the capabilities and limitations of processing tools. Currently, the recommended lead system, although empirical, is based on an international consensus advocating the use of three pseudo-orthogonal leads positioned according to the scheme shown in figure I.14:

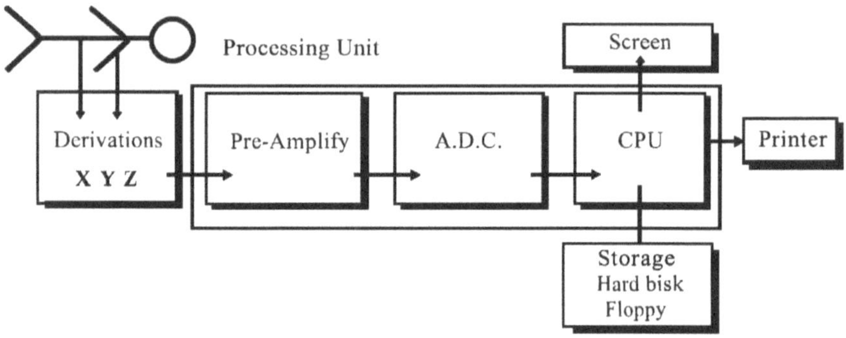

FigureI.14 : Overview of the ECG acquisition chain. ADC: Analog-to-Digital Converter; CPU: Central Processing Unit.

I.6 Associated electronics:

I.6.1 Amplifier :

I.6.1.1 Definition:

As their name suggests, amplifiers amplify signals collected from the patient, but their role is not limited to this. They must also eliminate parasitic electrical signals produced by the electromagnetic environment. This is why the amplifiers used in clinical electrophysiology are differential [19].

I.6.1.2 Application of the differential amplifier in electrocardiogram (ECG):

The electrical signals of cardiac activity (a few microvolts) are collected by medical sensors (electrodes). The signals that are visualized are generally measured relative to a so-called reference electrode (figure I.15). This electrode is placed at a specific location on the body. The process, therefore, involves measuring the potential of each electrode relative to the reference electrode (which does not constitute ground). A differential measurement is thus required, and a differential amplifier is necessary [26].

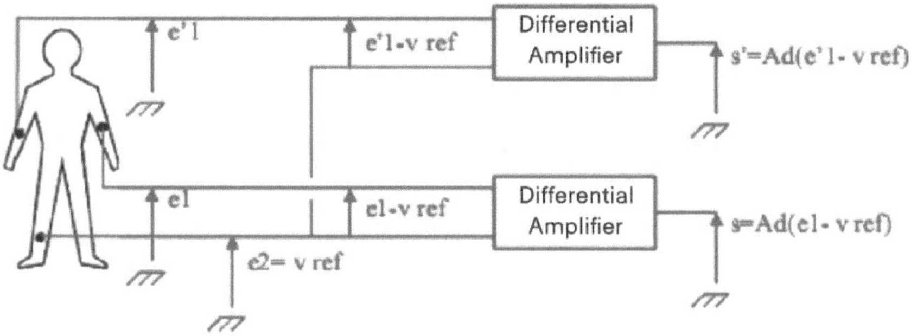

FigureI.15: application of the differential amplifier to the electrocardiogram (ECG).

I.6.2 ECG signal processing:

The recorder initially conditions the ECG signal collected by the electrodes. Subsequently, other primarily software-based processing can be applied to facilitate storage, remove noise and disturbances from the signal, analyze it, or contribute to automatic diagnosis.

I.6.2.1 Waves detection:

The detection of QRS complexes is of paramount importance in the automatic analysis of the ECG signal. Once the QRS complexes are identified and their positions are located, it becomes easy to evaluate other signal parameters, such as the duration of the cardiac cycle and the ST segment. Automatic QRS complex detection is a difficult task because the morphology of these complexes varies from one individual to another and even within the same individual from one cycle to another. Furthermore, other signal waves, such as P and T waves, and even disturbances of various origins, share characteristics with QRS complexes. Most detection algorithms proceed in two steps: the first step involves passing the signal through a bandpass filter that eliminates noise and P and T waves; the signal then undergoes a nonlinear transformation, for example, differentiation to identify steep slopes around the R wave, and squaring to quantify the energy of the QRS complexes. The second step consists of decision-making based on thresholding criteria [27, 28].

I.6.2.2 Filtering :

The ECG signal covers the frequency range 0.05–80 Hz, and the recorder circuits' bandwidth must be only slightly greater than that of the signal to achieve an optimal signal-to-noise ratio. The last stage of the recorder incorporates a bandpass filter that isolates the ECG signal according to these frequency characteristics. The recorder also includes a 50 Hz rejection filter that eliminates mains interference.

Other filtering operations aim to minimize the effects of interference artifacts caused by temporary poor electrode or cable contact, or by muscle movement. Figure I.16 establishes the spectral occupancy of the ECG signal, the QRS complex, and the P and T waves; the spectra of muscle noise and motion artifacts are also shown. Figure I.17 illustrates the effect of filtering on a noise-corrupted ECG [34].

The results produced by analog filters are insufficient. Much research is still dedicated to finding algorithms for purifying digitized ECG signals. Recent filtering algorithms are based on wavelet transformation.

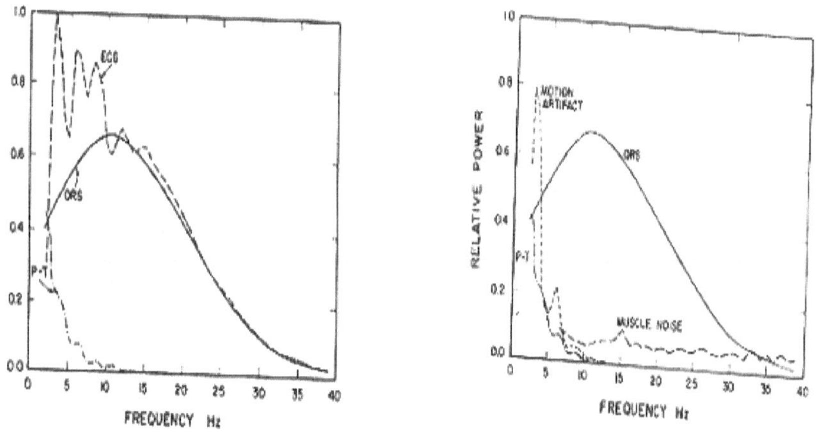

Figure I.16 : Spectral congestion of the ECG and certain characteristic waves.

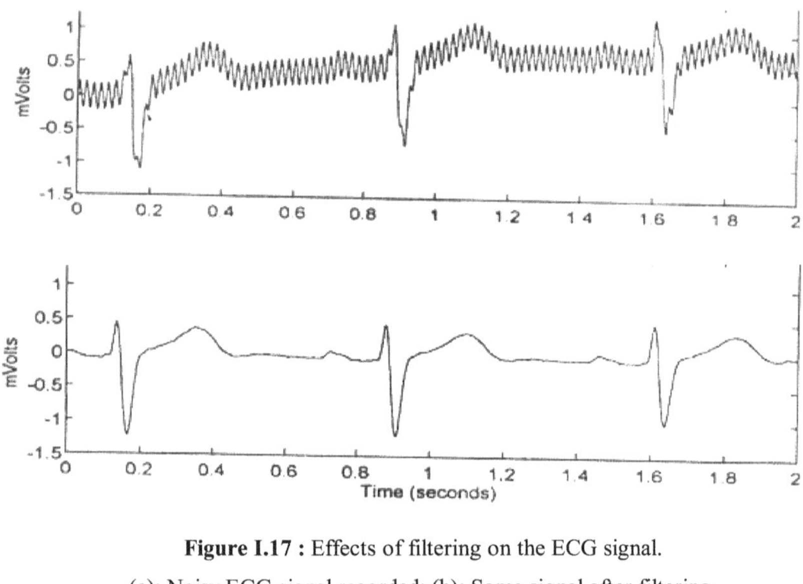

Figure I.17 : Effects of filtering on the ECG signal.

(a): Noisy ECG signal recorded; (b): Same signal after filtering.

Filtering is also an essential preliminary step for most QRS complex detection algorithms. The same applies to late recognition of potential.

On the left: ECG spectra, QRS complexes, P and T waves, and on the right: spectra of QRS, P and T waves, as well as muscle and motion artifacts.

I.7 Conclusion:

The measurement of ECG signals and amplification is established, unlike signal processing. In this area, the advent of digital technology has allowed us to enter a new dimension, thanks to its power and ease of use. Indeed, computers have brought speed, versatility, storage, and communication capabilities, and this is why an ECG acquisition system has been developed using the PIC16F877 microcontroller, which converts the signal and applies it, as we will detail in the following chapters.

Chapter 2

Architecture and Functional Overview of the PIC16F877 Microcontroller

II.1 Introduction:

The purpose of this basic knowledge is to understand the additional tools we need in this thesis, for example, the necessity and function of ports (A, E, C) or the oscillator, the definition of MCLR, etc. This knowledge relates to the properties and information for the integrated circuit known as the PIC16F877, which is simply a computer on a chip. It is one of the most important developments in electronics since the invention of the microprocessor itself. It is essential for the operation of devices such as mobile phones, DVD players, video cameras, etc. The small LCD screen is a good indicator of the presence of a MCU (microcontroller unit) – it needs a programmed device to control it. This chip sometimes works with other chips, but often it works on its own. The MCU provides the key element in the vast array of small, programmable devices that are commonplace today [35, 36].

II.2 Architecture of VON NEUMANN and HARVARD :

These microcontrollers deserve an introduction. Indeed, they were designed on a so-called **HARVARD (RISC)** architecture rather than on a **VON NEUMANN (COMPLEX)** model [37], as shown in Figures II.1 and II. 2.

The VON NEUMANN architecture used by most current microcontrollers (Intel 80XX, Motorola HC05, HC08, and HC11, or ZILOG Z80) is based on a single data bus. This bus carries both instructions and data. The HARVARD architecture used by PICS microcontrollers is based on two data buses. One bus is used for data and the other for instructions [38].

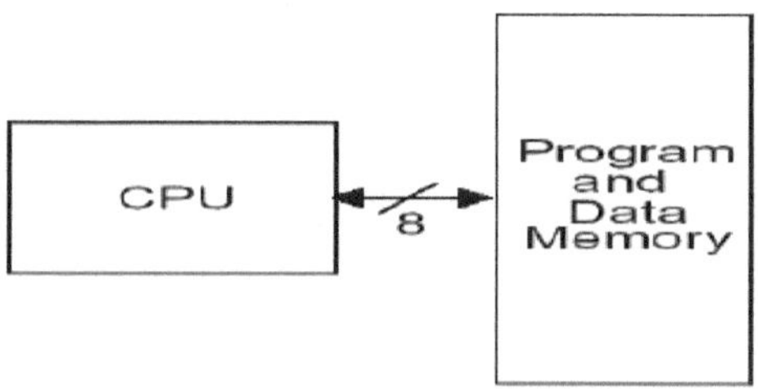

Figure II.1 : VON NEUMANN architecture.

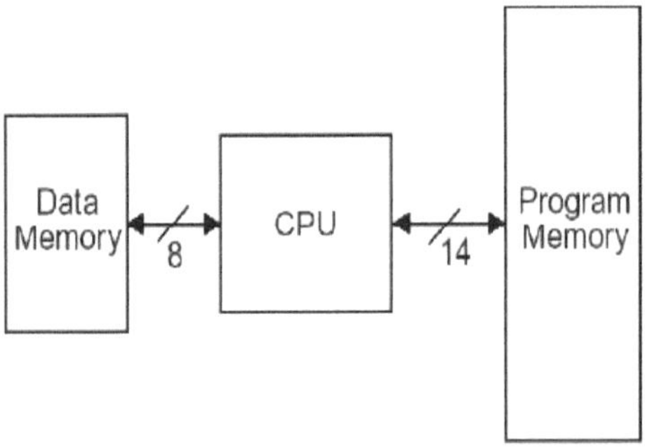

Figure II.2 : HARVARD architecture.

II.3 Difference between a processor and a microcontroller :

The microcontroller contains the same main components as any computer system:

- Processor.
- Memory.
- Input / Output.

In a PC, these chips are supplied as separate components connected via buses on a printed circuit board, but under the control of the microprocessor (CPU). A bus is a set of parallel lines that carry data and are shared by peripherals. The system can be designed to suit a particular need, with the type of processor, memory size, and selection of input/output (I/O) devices tailored to the system requirements[39].

In a microcontroller, all these elements are on a single chip. This means that the microcontroller for a particular application must be selected from the available options based on the requirements. In a given circuit, the microcontroller typically has a single dedicated function (unlike a PC); such systems are described as embedded applications [40].

II.4 Definition of PIC:

Microcontrollers are now used in most consumer and professional applications; there are several families of them. In the 1990s, the American company Microchip Technology developed a CMOS microcontroller: the PIC (Peripheral Interface Controller). This component, still widely used today, represents a compromise between ease of use, speed, and cost [41, 42].

PIC microcontrollers come in several versions:

* EPROMs that can be erased by a source of ultraviolet radiation
* OTPROMs are programmable only once
* EEPROMs and flash EPROMs that are electrically erasable.

II.4.1 Classification of microchip PICs:

Currently, MICROCHIP models are classified into three prominent families, each containing several references [41]. These families are:

* Baseline: instructions are coded on 12 bits.
* Mid-line: instructions are coded on 14 bits.
* High-end: instructions are coded on 16 bits.

II.4.2 Identification of PICs:

A PIC is generally identified by a reference of the following form: xx(L)XXyy-zz.

xx: component family, currently "12, 4, 16, 17, and 18".

L: greater tolerance of the voltage range.

XX: type of program

C: EPROM or EEPROM

F: flash

yy: identifier

zz: maximum speed of the pilot quartz.

II.4.3 comparative Table of PICs:

There are many types of PIC, in table II.1 we show a comparative Table of the hexadecimal values of PICs:

	Prog Mem in bytes	RAM in bytes	EEPROM in bytes	Fmax in MHz	I / O	Package
12C508	512x12	25	-	4	6	8 pins
16C72A	2048x14	128	-	20	22	28 pins
16F84	1024x14	68	64	20	13	18 pins
16F628	2028x14	224	128	20	16	18 pins
16F876	8192x14	368	256	20	22	28 pins
16F877	8192x14	368	256	20	33	40 pins

Table II.1: Comparative table of PICs [43].

II.5 Internal architecture of the PIC 16F877:

The PIC 16F877 is an integrated circuit contained in a package called "DIL 40"; it has 40 pins, 20 on each side. The pins are virtually numbered from 1 to 40. The first pin is located in the corner to the left of the locating notch[44, 45].

II.5.1 Pins of the µC 16F877:

II.5.1.1 MCLR :

This pin is used to initialize the microcontroller(figure II.3 and figure II.4). The microcontroller has several **RESET** points:

• **POR (POWER ON RESET) Mise sous tension**: A rising edge on **MCLR** triggers the microcontroller initialization. The time required is a minimum of 72ms and a maximum of 72ms + 1024 * Tosc. The microcontroller has an internal level detection circuit; when the **VDD** voltage is between 1.2V and 1.7V, it starts an initialization procedure [46].

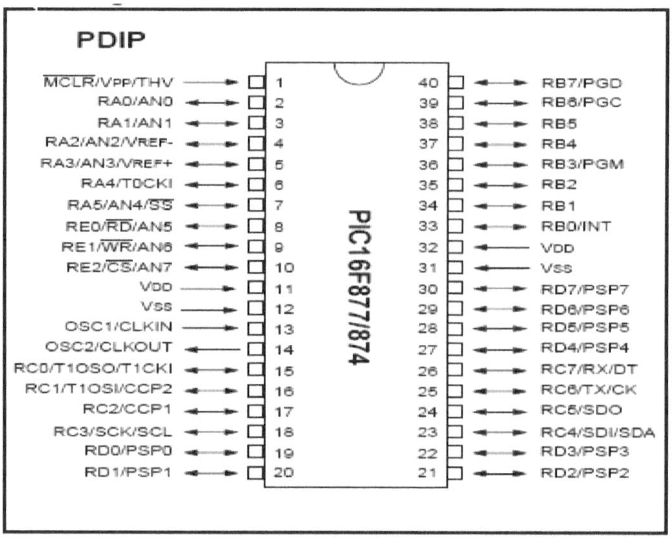

Figure II.3: Pins of 16F877.

This pin can be connected to **VDD** if an external **RESET** is not required. However, if a reset button is desired, a simple RC network can be wired to the **MCLR** pin.

• **EXTERNAL RESET (Mise à l'état bas de MCLR):** External reset. A low level must be applied to the RESET input for at least 2µS for the initialization to take effect.

• **WDT:** Watchdog timer. If the WDT reaches the end of the watch time without having been refreshed, the microcontroller will be initialized.

• **BOR:** Watchdog timer. If the VDD voltage drops below 4V for at least 100µS, the microcontroller may generate a RESET.

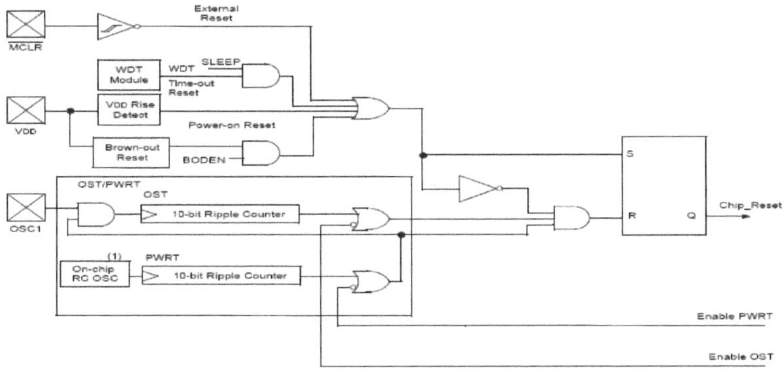

Figure II.4 : Structural diagram of the MCLR pin wiring.

II.5.1.2 Oscillator (OSC1 and OSC2 or CLKIN and CLOUT) :

Figure II.5 indicates that the internal oscillator functions of these pins allow the PIC's [47].

- **A quartz or ceramic resonator.**

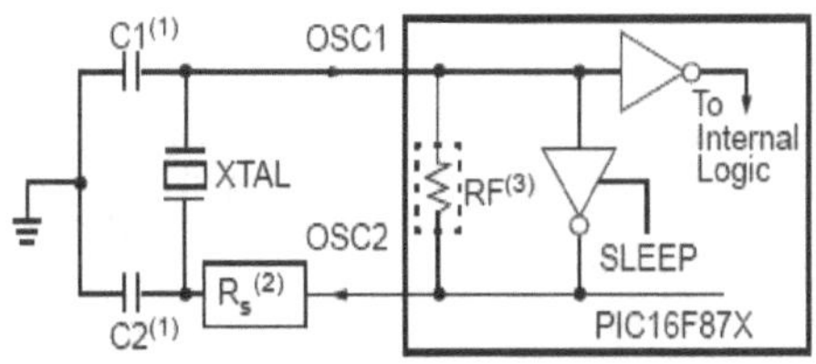

Figure II.5 : A quartz or ceramic resonator.

The RF varies with the Crystal chosen, and in both tables II.2 and II.3. There are recommended values of C1 and C2.

Ranges Tested:			
Mode	Freq.	OSC1	OSC2
XT	455 kHz 2.0 MHz 4.0 MHz	68 - 100 pF 15 - 68 pF 15 - 68 pF	68 - 100 pF 15 - 68 pF 15 - 68 pF
HS	8.0 MHz 16.0 MHz	10 - 68 pF 10 - 22 pF	10 - 68 pF 10 - 22 pF

Table II.2: Ranges tested of oscillator.

Osc Type	Crystal Freq.	Cap. Range C1	Cap. Range C2
LP	32 kHz	33 pF	33 pF
	200 kHz	15 pF	15 pF
XT	200 kHz	47-68 pF	47-68 pF
	1 MHz	15 pF	15 pF
	4 MHz	15 pF	15 pF
HS	4 MHz	15 pF	15 pF
	8 MHz	15-33 pF	15-33 pF
	20 MHz	15-33 pF	15-33 pF

Table II.3: Capacities of condensers of oscillation.

•**An external oscillator.**

It is one of the essential components of a microcontroller and its architecture shown in figure II.6

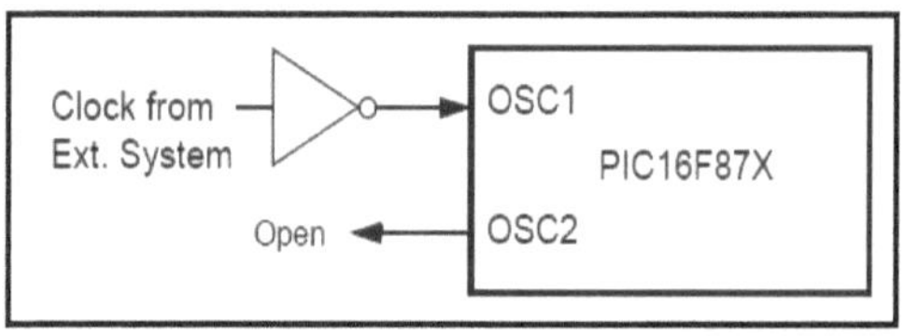

Figure II.6 : An external oscillator.

•**An RC network.**

The circuit network of PIC 16F87X contains many electronic components, as explained in figure II.7

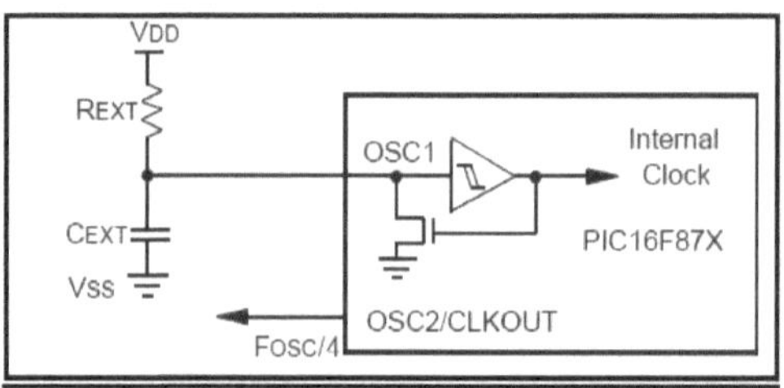

Figure II.7 : An RC network and recommended values:
3 k REXT 100 kΩ CEXT > 20 pF.

II.5.1.3 power supply (VDD et VSS):

These are the circuit's power supply pins [48]. The voltages that can be applied are:]

•From 4.5V to 6V for the standard F range.

•From 2 to 6V for the extended LF range.

The current drawn can range from 1 μA to 10 mA. The consumption per μC will depend on:

- The supply voltage.
- Internal frequency.
- The operating mode.

Furthermore, these terminals must be decoupled by two capacitors:

- **1μF** electrolytic.
- **10nF** ceramic.

II.5.2 CENTRAL UNIT :

II.5.2.1 Memory organization:

Since PICs use one bus for instructions and one bus for data, two memory planes must be considered, one for instructions and the other for data, as well as the internal registers [49].

II.5.2.1.1 Memory plan for instructions (program code):

The memory layout (figure II.8) is linear; addresses range from 0000h to 1FFFh (8k words of 14 bits), with each page containing 2k words. Note that the reset vector is fixed at 0000h. The PICs have only one interrupt vector at 0004h. During an interrupt, the associated subroutine must determine which device requested the interrupt.

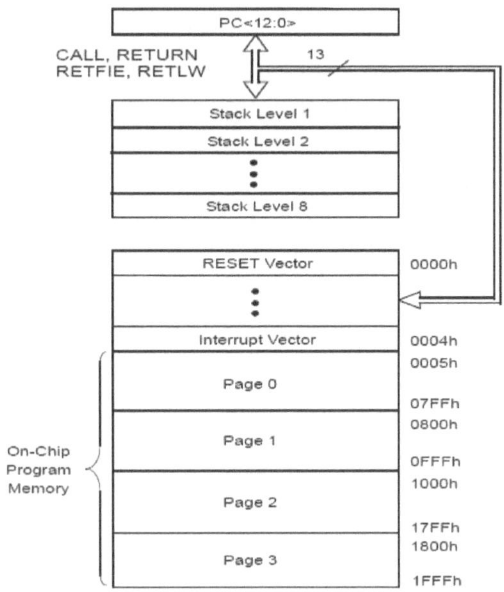

Figure II.8 : Memory plan for instructions (Program code).

The stack used by subroutines is not located in data memory, as with conventional microcontrollers, but in program memory. These stacks are used when subroutines are called; you cannot nest more than eight subroutines (which is already a lot!).

II.5.2.1.2 Data Memory Plan and Internal Registers (SFR):

The memory plan for data and internal registers is divided into four zones or banks of 128 bytes; to access a zone, you must set bits RP0 (bit 5) and RP1 (bit 6) of the STATUS register as indicated in table II.4. [50-52]

RP1 : RP0		Selected area (BANK)
0	0	From 00h to 7Fh: BANK 0
0	1	From 80h to FFh : BANK 1
1	0	From 100h to 17Fh : BANK 2
1	1	From 180h to 1FFh : BANK 3

Table II.4 : State registry configuration.

Figure II.9 shows the "General Purpose Registers" as nothing more than memory locations for storing data.

- The Work Register: It is called the 8-bit W: WORKING accumulator; PIC microcontrollers use this register to perform all kinds of calculations.

- **The 8-bit STATUS register**:

In Figure II.10, we show the different bits of the STATUS register, where each bit has a specific meaning. It is primarily used for testing-related tasks.

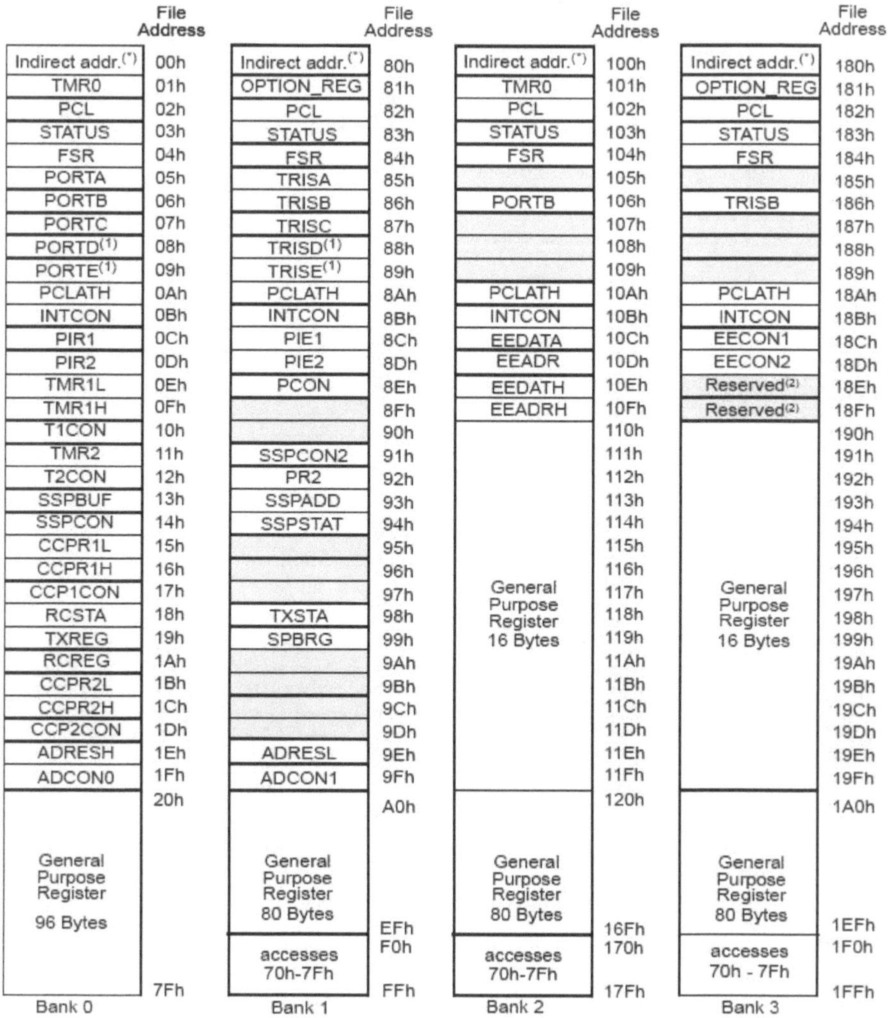

: Unimplemented data memory locations, read as '0'.

*: Not a physical register.

Note1: These registers are not implemented on the PIC16F876.

2: These registers are reserved, maintain these registers clear.

Figure II.9 : Special function registers.

R/W-0	R/W-0	R/W-0	R-1	R-1	R/W-x	R/W-x	R/W-x
IRP	RP1	RP0	$\overline{TO}$	$\overline{PD}$	Z	DC	C
bit 7							bit 0

bit 7 IRP: Register Bank Select bit (used for indirect addressing)
1 = Bank 2, 3 (100h - 1FFh)
0 = Bank 0, 1 (00h - FFh)

bit 6-5 RP1:RP0: Register Bank Select bits (used for direct addressing)
11 = Bank 3 (180h - 1FFh)
10 = Bank 2 (100h - 17Fh)
01 = Bank 1 (80h - FFh)
00 = Bank 0 (00h - 7Fh)
Each bank is 128 bytes

bit 4 $\overline{TO}$: Time-out bit
1 = After power-up, CLRWDT instruction, or SLEEP instruction
0 = A WDT time-out occurred

bit 3 $\overline{PD}$: Power-down bit
1 = After power-up or by the CLRWDT instruction
0 = By execution of the SLEEP instruction

bit 2 Z: Zero bit
1 = The result of an arithmetic or logic operation is zero
0 = The result of an arithmetic or logic operation is not zero

bit 1 DC: Digit carry/$\overline{borrow}$ bit (ADDWF, ADDLW, SUBLW, SUBWF instructions)
(for $\overline{borrow}$, the polarity is reversed)
1 = A carry-out from the 4th low order bit of the result occurred
0 = No carry-out from the 4th low order bit of the result

bit 0 C: Carry/$\overline{borrow}$ bit (ADDWF, ADDLW, SUBLW, SUBWF instructions)
1 = A carry-out from the Most Significant bit of the result occurred
0 = No carry-out from the Most Significant bit of the result occurred

Note: For $\overline{borrow}$, the polarity is reversed. A subtraction is executed by adding the two's complement of the second operand. For rotate (RRF, RLF) instructions, this bit is loaded with either the high, or low order bit of the source register.

Legend:			
R = Readable bit	W = Writable bit	U = Unimplemented bit, read as '0'	
- n = Value at POR	'1' = Bit is set	'0' = Bit is cleared	x = Bit is unknown

Figure II.10 : The state register.

•The "PCL" and "PCLATH" registers:

The PCL register contains the PC's least significant address (8 bits). The PCLATH register contains the PC's most significant address (5 bits) (figure II.11).

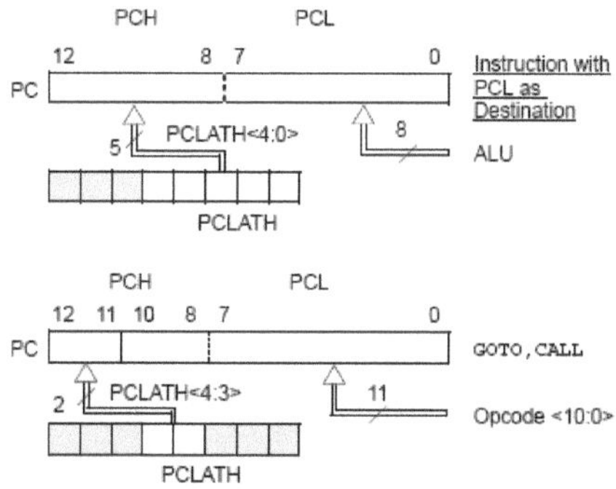

Figure II.11: PC loading in different cases.

•PORTx and TRISx registers :

The PORTx register determines the logical state of the outputs when PORTx or specific lines of PORTx are configured as outputs.

The TRISx register is the direction register. It determines whether PORTx or specific port lines are input or output. Writing a logic one corresponds to an input (1 for Input) and writing a logic 0 corresponds to an output (0 for Output).

Upon RESET, all port lines are configured as inputs.

•FSR and INDF registers:

In PIC microcontrollers, the available indirect addressing pointer is FSR. Using direct addressing, the address of the register to be accessed can be written to FSR. Access to the register whose address is contained in FSR is done using the INDF register. The figure illustrates the working method.

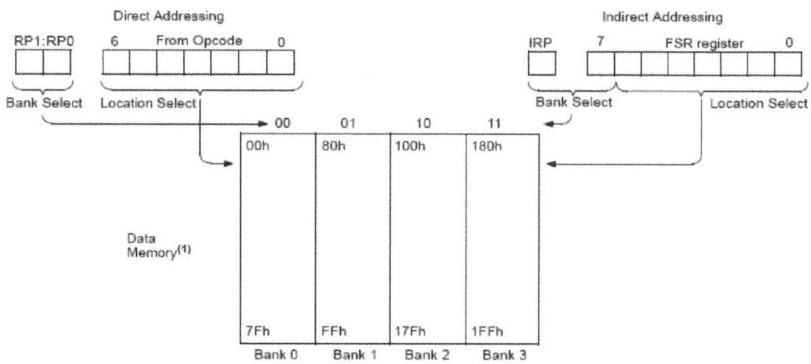

Figure II.12 : Direct/Indirect Addressing.

•The timer0 registers:

The configuration of timer 0 (figure II.13)goes through the registers TMR0 (address 01h), OPTION_REG (address 81h page 1), and INTCON (address 0Bh: all pages).

OPTION_REG REGISTER (@ 81h or 181h)

R/W-1	R/W-1	R/W-1	R/W-1	R/W-1	R/W-1	R/W-1	R/W-1
$\overline{RBPU}$	INTEDG	T0CS	T0SE	PSA	PS2	PS1	PS0
bit 7							bit 0

bit 7 $\overline{RBPU}$

bit 6 INTEDG

bit 5 T0CS: TMR0 Clock Source Select bit
1 = Transition on T0CKI pin
0 = Internal instruction cycle clock (CLKOUT)

bit 4 T0SE: TMR0 Source Edge Select bit
1 = Increment on high-to-low transition on T0CKI pin
0 = Increment on low-to-high transition on T0CKI pin

bit 3 PSA: Prescaler Assignment bit
1 = Prescaler is assigned to the WDT
0 = Prescaler is assigned to the Timer0 module

bit 2-0 PS2:PS0: Prescaler Rate Select bits

Bit Value	TMR0 Rate	WDT Rate
000	1 : 2	1 : 1
001	1 : 4	1 : 2
010	1 : 8	1 : 4
011	1 : 16	1 : 8
100	1 : 32	1 : 16
101	1 : 64	1 : 32
110	1 : 128	1 : 64
111	1 : 256	1 : 128

Legend:		
R = Readable bit	W = Writable bit	U = Unimplemented bit, read as '0'
- n = Value at POR	'1' = Bit is set	'0' = Bit is cleared x = Bit is unknown

REGISTERS ASSOCIATED WITH TIMER0

Address	Name	Bit 7	Bit 6	Bit 5	Bit 4	Bit 3	Bit 2	Bit 1	Bit 0	Value on: POR, BOR	Value on all other RESETS
01h,101h	TMR0	Timer0 Module's Register								xxxx xxxx	uuuu uuuu
0Bh,8Bh, 10Bh,18Bh	INTCON	GIE	PEIE	T0IE	INTE	RBIE	T0IF	INTF	RBIF	0000 000x	0000 000u
81h,181h	OPTION_REG	RBPU	INTEDG	T0CS	T0SE	PSA	PS2	PS1	PS0	1111 1111	1111 1111

Legend: x = unknown, u = unchanged, - = unimplemented locations read as '0'.
Shaded cells are not used by Timer0.

Figure II.13 : The configuration of the timer0 registers and the relationship between them.

•The timer1 registers:

Figure II.14 shows the configuration of timer 1, which goes through the registers: PIR1 (address 0Ch), PIE1 (address 8Ch page 1), the registers TMR1L and TMR1H (addresses 0Eh and 0Fh), T1CON (address 10h page 0), and INTCON (address 0Bh: all pages).

T1CON: TIMER1 CONTROL REGISTER (ADDRESS 10h)

U-0	U-0	R/W-0	R/W-0	R/W-0	R/W-0	R/W-0	R/W-0
—	—	T1CKPS1	T1CKPS0	T1OSCEN	$\overline{\text{T1SYNC}}$	TMR1CS	TMR1ON
bit 7							bit 0

bit 7-6 Unimplemented: Read as '0'

bit 5-4 T1CKPS1:T1CKPS0: Timer1 Input Clock Prescale Select bits
11 = 1:8 Prescale value
10 = 1:4 Prescale value
01 = 1:2 Prescale value
00 = 1:1 Prescale value

bit 3 T1OSCEN: Timer1 Oscillator Enable Control bit
1 = Oscillator is enabled
0 = Oscillator is shut-off (the oscillator inverter is turned off to eliminate power drain)

bit 2 $\overline{\text{T1SYNC}}$: Timer1 External Clock Input Synchronization Control bit
When TMR1CS = 1:
1 = Do not synchronize external clock input
0 = Synchronize external clock input
When TMR1CS = 0:
This bit is ignored. Timer1 uses the internal clock when TMR1CS = 0.

bit 1 TMR1CS: Timer1 Clock Source Select bit
1 = External clock from pin RC0/T1OSO/T1CKI (on the rising edge)
0 = Internal clock (Fosc/4)

bit 0 TMR1ON: Timer1 On bit
1 = Enables Timer1
0 = Stops Timer1

Legend:			
R = Readable bit	W = Writable bit	U = Unimplemented bit, read as '0'	
- n = Value at POR	'1' = Bit is set	'0' = Bit is cleared	x = Bit is unknown

REGISTERS ASSOCIATED WITH TIMER1 AS A TIMER/COUNTER

Address	Name	Bit 7	Bit 6	Bit 5	Bit 4	Bit 3	Bit 2	Bit 1	Bit 0	Value on: POR, BOR	Value on all other RESETS
0Bh,8Bh, 10Bh, 18Bh	INTCON	GIE	PEIE	T0IE	INTE	RBIE	T0IF	INTF	RBIF	0000 000x	0000 000u
0Ch	PIR1	PSPIF(1)	ADIF	RCIF	TXIF	SSPIF	CCP1IF	TMR2IF	TMR1IF	0000 0000	0000 0000
8Ch	PIE1	PSPIE(1)	ADIE	RCIE	TXIE	SSPIE	CCP1IE	TMR2IE	TMR1IE	0000 0000	0000 0000
0Eh	TMR1L	Holding Register for the Least Significant Byte of the 16-bit TMR1 Register								xxxx xxxx	uuuu uuuu
0Fh	TMR1H	Holding Register for the Most Significant Byte of the 16-bit TMR1 Register								xxxx xxxx	uuuu uuuu
10h	T1CON	—	—	T1CKPS1	T1CKPS0	T1OSCEN	$\overline{\text{T1SYNC}}$	TMR1CS	TMR1ON	--00 0000	-- uu uuuu

Legend: x = unknown, u = unchanged, - = unimplemented, read as '0'. Shaded cells are not used by the Timer1 module.
Note 1: Bits PSPIE and PSPIF are reserved on the PIC16F873/876; always maintain these bits clear.

Figure II.14 : The configuration of the timer1 registers and the relationship between them.

•The timer2 registers:

The configuration of timer 2 goes through the registers: PIR1 (address 0Ch), PIE1 (address 8Ch page 1), TMR2 (address 11h page 0), PR2 (address 92h page 1), T2CON (address 12h page 0), and INTCON (address 0Bh: all pages) as shown in figure II.15.

T2CON: TIMER2 CONTROL REGISTER (ADDRESS 12h)

U-0	R/W-0	R/W-0	R/W-0	R/W-0	R/W-0	R/W-0	R/W-0
—	TOUTPS3	TOUTPS2	TOUTPS1	TOUTPS0	TMR2ON	T2CKPS1	T2CKPS0
bit 7							bit 0

bit 7 Unimplemented: Read as '0'

bit 6-3 TOUTPS3:TOUTPS0: Timer2 Output Postscale Select bits
0000 = 1:1 Postscale
0001 = 1:2 Postscale
0010 = 1:3 Postscale
□
□
□
1111 = 1:16 Postscale

bit 2 TMR2ON: Timer2 On bit
1 = Timer2 is on
0 = Timer2 is off

bit 1-0 T2CKPS1:T2CKPS0: Timer2 Clock Prescale Select bits
00 = Prescaler is 1
01 = Prescaler is 4
1x = Prescaler is 16

Legend:			
R = Readable bit	W = Writable bit	U = Unimplemented bit, read as '0'	
- n = Value at POR	'1' = Bit is set	'0' = Bit is cleared	x = Bit is unknown

Figure II.15 : Configuring the timer2 registers.

•Interrupt registers:

The registers involved in interrupt handling are INTCON, PIR1, PIR2, PIE1, PIE2, and PCON. Interrupts are external signals that force the MCU to suspend its current process and perform an interrupt service routine (ISR). An interrupt can be generated in various ways, but in the PIC, it always jumps to program address 004. If more than one interrupt source is active, the interrupt source must be detected and the corresponding ISR selected.

•CCP1 and CCP2 module registers (CCP: Capture Compare PWM):

These modules are associated with the RC2/CCP1 and RC1/T1OSI/CCP2 pins. Depending on the selected mode, timers 1 or 2 will be used. Capture and Comparison modes use timer 1, while PWM mode uses timer 2. It stores the value of timer 1 in the CPP1R1H: CPP1R1L or CPP1R2H: CPP1R2L registers when an event occurs on one of the CCP1 or CCP2 pins.

•The registers of the Conversion Module:

The converter works with the SFR registers: ADCON1, ADCON0.

•The USART module registers :

The special function registers of this module are: TXSTA, SPBRG, RCSTA, TXREG, and RCREG.

II.5.3 INPUT/OUTPUT PORTS:

The PIC 16F877 contains five groups of inputs/outputs (A, B, C, D), which are summarized in table II.5: [53]

	Bit	Brooches	Other functions	Bit	By default
PORTA	6	RA0-RA5	Analog inputs	0,1,2,3	Analog inputs
			Timer0 clock input	4	
			Selected serial port slave input	5	
PORTB	8	RB0-RB7	External interruption	0	E/S Digital
			Low voltage programming input.	3	
			Serial programming.	6,7	
			in-situ debugger.	6,7	
PORTC	8	RC0-RC7	Timer1 Clock I/O	0,1	E/S Digital
			Capture/comparison/PWM	1,2	
			SPI, I2C synchronous clock/data	3,4,5	
			USART asynchronous clock / data	6,7	
PORTD	8	RD0-RD7	data port slave input/output	0-7	E/S Digital
PORTE	3	RE0-RE2	Analog inputs	0,1,2	Analog inputs

Table II.5 : input/output of PIC 16F877.

II.6 Conclusion:

We have provided a general overview of this type of microcontroller, then defined the PIC family, specifically the 16F877.

In this chapter, the 16F877 microcontroller can serve as an analog-to-digital converter for our system, which is the most essential part of our project for building the ECG device. The circuit can be used for serial communication with the UART module.

To make this control unit work, it needs to be programmed and adapted to a programming compiler.

Chapter 3

Signal Conversion and Communication Systems for ECG Acquisition

III.1 Introduction:

This chapter generally discusses serial communication and data conversion using the RS232 interface. Serial communication involves transmitting information by dividing it into several fixed-size segments (each segment is the number of available communication lines). This contrasts with parallel communication, where the data to be transmitted does not need to be divided before transmission (because there are at least as many communication lines as there are bits to communicate). Interfacing an ECG signal is an example of serial communication [54, 55].

III.2 The conversion:

III.2.1 Data conversion databases:

As predicted by Moore's Law in 1964, digital computing power has increased exponentially, while costs have also increased. For example, our PIC16F877 can replace several circuits. With this increase in computing power, many applications usually performed with analog circuitry are finding a new life in the digital world [56, 57]. However, the real world will always remain fundamentally analog. To bring the digital processing capabilities of the PIC16F877 and its advantages to real-world applications, the analog signal of interest must be converted to a format the PIC16F877 can understand. This process is performed by an analog-to-digital converter (ADC). After processing by the PIC16F877, the resulting digital information stream must be converted back to an analog form by a digital-to-analog converter (DAC). Once again, with an analog signal, the information can be perceived by the human senses, most often sight or hearing. An illustration of this stream is shown in figure III.1.

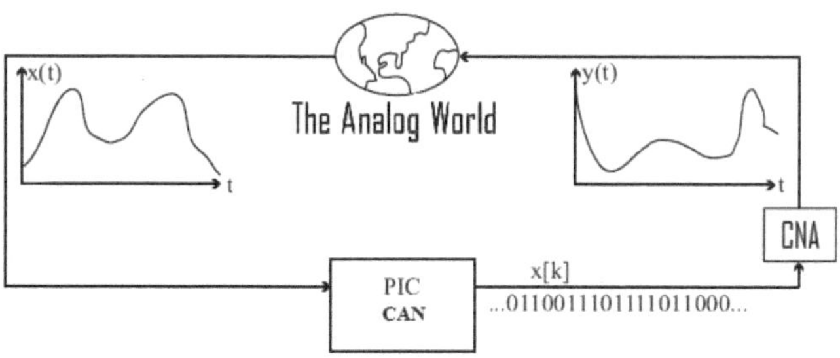

Figure III.1 : A typical use of a PIC is to connect via an interface to the "analog" world.

Converters are ubiquitous in computing systems. Many electronic products, including disk drives, camcorders, digital mobile phones, modems, computer sound cards, computer graphics adapters, and high-definition televisions, contain one or more data converters. Since ADCs are very useful and required by many small microprocessor applications, microprocessor architects often include an ADC as an integrated peripheral, and the designers of the PIC16F877 chip did just that.

From a programmer's perspective, the ADCs and DACs in Figure III.1 can be considered black boxes. That is, an ADC accepts an analog input of a certain quantity, typically a voltage, and provides n bits of digital output every fs seconds (where fs is the ADC's sampling frequency) that represent this analog input. For most purposes, this is a sufficient interpretation of data converters. However, understanding how data conversion is performed will help you understand the limitations on the operation of ADCs and DACs and should help you choose data converters for your specific application [58].

III.2.2 The PIC 16F877 conversion module :

The 16F87x uses an analog-to-digital converter that allows 10-bit sampling. The digital signal can therefore take on 1024 possible values. Figure III.2 indicates how to convert the quantities; their minimum and maximum values must be known. By default, PIC® microcontrollers treat the minimum value as their supply voltage (Vss), while the maximum value corresponds to the positive supply voltage (Vdd). However, other values can be used [59]. The methods used to convert analog quantities into digital amounts are at the heart of the PIC®.

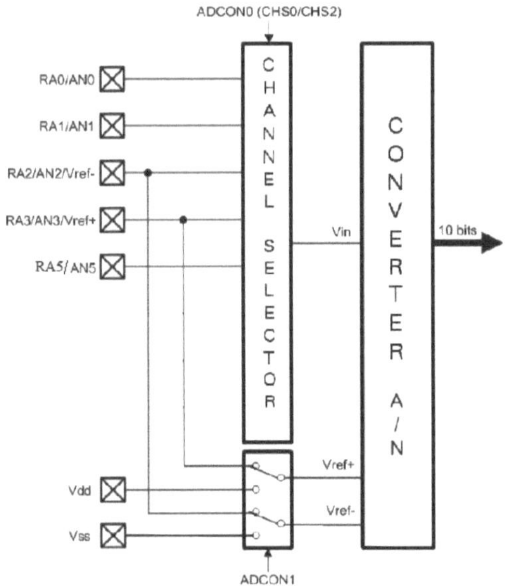

Figure III.2 : The conversion modulator.

The sequence for physically converting the analog quantity into a digital quantity within the PIC® is as follows:

- The PIC® connects the pin containing the voltage to be digitized to an internal capacitor, which charges via an internal resistor until it reaches the applied voltage.

- The pin is disconnected from the capacitor, and the latter is connected to the internal analog/digital converter.
- The PIC® carries out the conversion.

III.2.2.1 Acquisition time :

This is the time it takes for the internal capacitor to reach a voltage close to the target voltage. This charging occurs through an internal resistance and the resistance of the source connected to the pin. This time is incremented by the response time of the internal circuits and by a time that depends on temperature (the temperature coefficient). It is important to note that resistances increase with temperature, and therefore, the circuits' response times also increase [60].

So, if we assume: T_{acq} = total acquisition time, T_{amp} = circuit reaction time, T_c = capacitor charging time, and T_{coff}= time which depends on the temperature coefficient.

The formula is therefore :

$$T_{acq} = T_{amp} + T_c + T_{coff}$$

The reaction time T_{amp} is typically 2µs, so there is no problem there: Tamp = 2µs

The temperature coefficient is only necessary for temperatures above 25°C. In other cases, it is not relevant. This coefficient is typically 0.05 µs per °C above 25°C. This refers to the PIC® temperature, not the ambient temperature [61].

Therefore, this Tcoff time will be at least 0 (below 25°C) and at most (50-25)*0.05, or 1.25 µs. The PIC® temperature cannot, in fact, exceed 50°C.

First observation: to benefit from the maximum frequency, we must keep the PIC® below 25°C. The remaining factor is charging time. This charging time depends on the resistance placed in series with the capacitor. In fact, there are two resistances: the one in our signal source and the one inside the PIC®. It is recommended that the resistance of our source remain below 10 kΩ. The internal resistance of the PIC® is directly related to the supply voltage. The lower the voltage, the higher the resistance, and therefore the longer the charging time. So, again, to achieve high speeds, we will need to power the PIC® with the maximum supported voltage, which is currently 6V for the 16F876. The total internal resistance (composed of two internal resistors) varies from 6 kΩ at 6V to 12 kΩ at 3V, passing through 8 kΩ at 5V [62]. Furthermore, since the capacitor's charge also depends on the signal source's resistance, to increase our speed, we should also use a signal source with the lowest possible impedance (resistance).

Given that the internal capacitor has a value of 120 pF in current PIC® (16F876) devices, the formulas I gave you for calculating a capacitor's charging time remain valid. The formula for the capacitor charging time is:

$$T_c = C * (R_{interne} + R_{source}) * \ln(1/2047).$$

The 2047 comes from the fact that, to digitize with a precision of 1/2 bit, using a maximum value of 1023, the capacitor charge must be at least 2046/2047ths of the voltage to be measured. Since "C" is fixed and "ln(1/2047)" is also fixed, the constant will be:

$$C * \ln(1/2047) = 0{,}914895 * 10^{-9}$$

The formula, therefore, becomes:

$$T_c = 0{,}914895 * 10^{-9} * (R_{interne} + R_{source})$$

If we consider the worst-case scenario (3V voltage and 10kΩ source resistance), our charging time is: $T_c = 0{,}914895 * 10^{-9} * (10 * 10^3 + 12 * 10^3)$ and $T_{c\ maximal} = 20{,}12\ \mu s$

Now the most favorable case (voltage of 6V, and negligible source resistance):

$T_c = 0{,}914895 * 10^{-9} * (0 * 10^3 + 6 * 10^3)$ and $T_{c\ minimal} = 5{,}48\ \mu s$.

In the typical case, namely a supply voltage of 5V and a source resistance of 10 kΩ, we have: Tc = 0,914895 × 10^-9 × (10 × 10^3 + 8 × 10^3) and Tc typique = 16,46 µs.

Now we calculate the minimum, maximum, and typical times of the total acquisition time T_{acq}.

The worst-case scenario is: a temperature of 50°C and a maximum Tc, which gives:

$$T_{acq} = T_{amp} + T_{ac} + T_{coff}$$

$T_{acq\ maximum} = 2\mu s + 20{,}12\mu s + 1{,}25\mu s = 23{,}37\ \mu s$

The most favorable case, a temperature less than or equal to 25°C and a minimum Tc, gives us: $T_{acq\ minimum} = 2\ \mu s + 5{,}48\ \mu s = 7{,}48\ \mu s$.

III.2.2.2 The conversion:

After the Tacq time, the capacitor can be considered charged and ready to be connected to the analog-to-digital converter input. This connection takes approximately 100ns. Once the capacitor is connected and the voltage to be digitized is present at the converter's input, the converter must perform the conversion. The principle used is that of successive approximation [63-65].

The time required for the conversion is equal to the time required to convert one bit multiplied by the number of bits desired for the result. Regarding our PIC®, it is essential to know that converting a bit requires a time called Tad. This time is derived by dividing the central clock. The divisor can take the values 2, 8, or 32.

III.2.2.3 Compromise between speed and precision:

Increasing speed beyond these limits can only be achieved at the expense of accuracy in terms of the number of significant bits in the result. This increase also comes at the cost of greater programming difficulty. Here is the reasoning used. Suppose we accept limiting the precision to fewer than 10 bits and note that digitization is carried out starting with the most significant bits. In that case, we can say to ourselves: N bits are enough for me, we decide to stop the conversion after the bits in question have been digitized [66]. Therefore, we assume that to retain n significant bits, we would have:

• T_{acq} to charge the capacitor.

• T_{ad} for startup.

• n * T_{ad} for conversion.

So, for a conversion keeping 4 bits instead of 12, this would bring our total time to

$T_{acq} + T_{ad} + 4\ T_{ad}$, or $T_{acq} + 5\ T_{ad}$., in this case, 7 T_{ad} have been saved.

Unfortunately, it is not possible to stop the digitization process, as the results would not be recorded in the relevant registers. Therefore, this method requires improvement. While the digitization itself cannot be halted, the value of the divisor being digitized can be altered.

Let us assume the task is completed with a PIC® running at 20 MHz. Therefore, the division will be by 32. Each Tad is thus worth 32 Tosc. Our adequate conversion time, therefore, is usually 12 Tad, or 12 × 32 Tosc, or 19.2 µs. The need for four valid bits leads to the following operations:

• T_{acq} to charge the capacitor.

• T_{ad} for startup = 32 Tosc.

• n * T_{ad} for conversion = n * 32 Tosc.

• Change the divisor to pass to 2.

• (11 – N) T_{ad} remaining for the rest of the conversion (which will be incorrect), therefore (11-N) * 2Tosc

(11-4) * 30 Tosc = 210 Tosc, or 210 * 50ns = 10.5µs

The general formula for the conversion time to digitize to n bits is therefore: Conversion time on n bits = Tad + N * Tad + (11-N) (2Tosc)

In other words:

Conversion time = (N+1) T_{ad} + (11-N) (2Tosc). This will give us, for our 4-bit conversion for a 20MHz PIC®:

Conversion time = (4+1) * 1.6µs + (11-4) * 100ns = 8µs + 0.7µs = 8.7µs. To this we add 2Tad, or 200ns, which gives us a time separating two samples of 16.38µs, so the sampling frequency will be: Sampling frequency = 1/16.38 µs = 61050 Hz, with 16 possible values. The second helpful tip is that if we do not want to use all the bits, we do not need to fully charge the capacitor. We can therefore recalculate the capacitor's charge to reduce T_{acq}, thereby increasing the sampling frequency slightly.

Note that using these techniques will always give us a 10-bit result, but only the first n bits will still represent a usable value. The other bits must therefore be ignored [67]. Of course, it will be up to us to determine the time between the start of the conversion process and the modification of the divisor. This modification will logically occur at a time Tad (n+1) after the conversion is initiated.

Since Tad is an exact multiple (usually 32) of the main oscillator, a simple loop or a timer will suffice.

For example, to digitize to 6 bits with a 20MHz PIC®:

- We launch the acquisition.
- After the Tacq time, we start the conversion.
- After the time (6+1) * Tad, we update the divisor and wait for the conversion to finish.

The waiting time (6+1) Tad is 7 * Tad, therefore, 7 * 32 Tosc. Knowing that the execution time of an instruction is Tosc/4, the waiting time will be 7 * 32 / 4 = 56 instruction cycles.

III.2.2.4 The values represented:

For a 10-bit scan, the resulting numerical value will be:

$V_{al} = ((V_e - V_{REF-}) / (V_{REF+} - V_{REF-})) * 1023)$

We suppose :

V_{REF-} : Minimum analog voltage (negative reference) and V_{REF+} : Maximum analog voltage (positive reference)

V_e: Input voltage to be digitized, and V_{al}: numerical value obtained using 10 bits. Conversely, the typical value that was digitized corresponds to a voltage of:

$V_e = ((V_{al} /1023) * (V_{REF+} - V_{REF-})) + V_{REF-}$

If we use a negative reference voltage of 0V, meaning that the negative voltage reference is actually Vss, we obtain two simplified formulas:

$V_{al} = (V_e / V_{REF+}) * 1023)$ and $V_e = (V_{al} /1023) * V_{REF+}$

If: $V_{REF+} = 5V$, $V_{REF-} = 0V$ then :

$V_{al} = (V_e / 5) * 1023)$ and $V_e = (V_{al} /1023) * 5$

III.2.2.5 Configuration :

- **The ADCON1 register:**

It allows us to choose one of the 16 configurations offered in figure III.3 and table III.1

ADCON1 REGISTER (ADDRESS 9Fh)

U-0	U-0	R/W-0	U-0	R/W-0	R/W-0	R/W-0	R/W-0
ADFM	—	—	—	PCFG3	PCFG2	PCFG1	PCFG0
bit 7							bit 0

bit 7 **ADFM:** A/D Result Format Select bit
1 = Right justified. 6 Most Significant bits of ADRESH are read as '0'.
0 = Left justified. 6 Least Significant bits of ADRESL are read as '0'.

bit 6-4 **Unimplemented:** Read as '0'

bit 3-0 **PCFG3:PCFG0:** A/D Port Configuration Control bits:

Figure III.3 : The ADCON register configuration.

PCFG3: PCFG0	AN7(1) RE2	AN6(1) RE1	AN5(1) RE0	AN4 RA5	AN3 RA3	AN2 RA2	AN1 RA1	AN0 RA0	VREF+	VREF-	CHAN/ Refs(2)
0000	A	A	A	A	A	A	A	A	VDD	Vss	8/0
0001	A	A	A	A	VREF+	A	A	A	RA3	Vss	7/1
0010	D	D	D	A	A	A	A	A	VDD	Vss	5/0
0011	D	D	D	A	VREF+	A	A	A	RA3	Vss	4/1
0100	D	D	D	D	A	D	A	A	VDD	Vss	3/0
0101	D	D	D	D	VREF+	D	A	A	RA3	Vss	2/1
011x	D	D	D	D	D	D	D	D	VDD	Vss	0/0
1000	A	A	A	A	VREF+	VREF-	A	A	RA3	RA2	6/2
1001	D	D	A	A	A	A	A	A	VDD	Vss	6/0
1010	D	D	A	A	VREF+	A	A	A	RA3	Vss	5/1
1011	D	D	A	A	VREF+	VREF-	A	A	RA3	RA2	4/2
1100	D	D	D	A	VREF+	VREF-	A	A	RA3	RA2	3/2
1101	D	D	D	D	VREF+	VREF-	A	A	RA3	RA2	2/2
1110	D	D	D	D	D	D	D	A	VDD	Vss	1/0
1111	D	D	D	D	VREF+	VREF-	D	A	RA3	RA2	1/2

A = Analog input D = Digital I/O

Table III.1 : 16 possible converter combinations.

- **The ADCON0 register::**

Table III.2 indicates the register allows setting of the conversion clock (ADCS1 and ADCS0 bits), the channel to be converted (CHS2, CHS1, and CHS0), and the ADON power-on bit. The clock selection is determined by the ADCS1 and ADCS0 bits, with the TAD time being at least 1.6 μs as represented in table III.3 and figure III.4 [68].

AD Clock Source (TAD)		Maximum Device Frequency
Operation	**ADCS1:ADCS0**	**Max.**
2Tosc	00	1.25 MHz
8Tosc	01	5 MHz
32Tosc	10	20 MHz
RC(1, 2, 3)	11	**(Note 1)**

Note 1: The RC source has a typical TAD time of 4 μs, but can vary between 2-6 μs.
2: When the device frequencies are greater than 1 MHz, the RC A/D conversion clock source is only recommended for SLEEP operation.
3: For extended voltage devices (LC), please refer to the Electrical Characteristics (Sections 15.1 and 15.2).

Table III.2 : The estimator's selection of ADCON0.

ADCON0 REGISTER (ADDRESS: 1Fh)

R/W-0	R/W-0	R/W-0	R/W-0	R/W-0	R/W-0	U-0	R/W-0
ADCS1	ADCS0	CHS2	CHS1	CHS0	GO/$\overline{\text{DONE}}$	—	ADON
bit 7							bit 0

bit 7-6 **ADCS1:ADCS0:** A/D Conversion Clock Select bits
00 = Fosc/2
01 = Fosc/8
10 = Fosc/32
11 = FRC (clock derived from the internal A/D module RC oscillator)

bit 5-3 **CHS2:CHS0**: Analog Channel Select bits
000 = channel 0, (RA0/AN0)
001 = channel 1, (RA1/AN1)
010 = channel 2, (RA2/AN2)
011 = channel 3, (RA3/AN3)
100 = channel 4, (RA5/AN4)
101 = channel 5, (RE0/AN5)(1)
110 = channel 6, (RE1/AN6)(1)
111 = channel 7, (RE2/AN7)(1)

bit 2 **GO/$\overline{\text{DONE}}$**: A/D Conversion Status bit
If ADON = 1:
1 = A/D conversion in progress (setting this bit starts the A/D conversion)
0 = A/D conversion not in progress (this bit is automatically cleared by hardware when the A/D conversion is complete)

bit 1 **Unimplemented**: Read as '0'

bit 0 **ADON**: A/D On bit
1 = A/D converter module is operating
0 = A/D converter module is shut-off and consumes no operating current

Note 1: These channels are not available on PIC16F873/876 devices.

Legend:		
R = Readable bit	W = Writable bit	U = Unimplemented bit, read as '0'
- n = Value at POR	'1' = Bit is set	'0' = Bit is cleared x = Bit is unknown

Figure III.4 : The ADCON0 register configuration.

CHS2	CHS1	CHS0	Channel	Pin
0	0	0	0	AN0/RA0
0	0	1	1	AN1/RA1
0	1	0	2	AN2/RA2
0	1	1	3	AN3/RA3
1	0	0	4	AN4/RA5
1	0	1	5	AN5/RE0 (Only for 16F877)
1	1	0	6	AN6/RE1 (Only for 16F877)
1	1	1	7	AN7/RE2 (Only for 16F877)

Table III.3 : Channels configuration of ADCON0.

III.3 Communication :

III.3.1 The principles of communication:

Each data transmission system requires:

- A data source (a transmitter or pilot line), which converts information into a form suitable for transmission over a link.
- A receiver that accepts the signal and converts it back into the original data.
- A communication link that carries signals. This can be a copper wire, a fiber-optic link, or a radio or satellite link.

Furthermore, the sender and receiver must be able to understand each other. This requires agreement on several factors. The most important are:

- The type of signage used.
- Definition of «logical 1» and «logical 0».
- The codes that represent the symbols.
- Maintain synchronization between the transmitter and receiver if this property is used.
- How the data flow is controlled so that the receiver is not overwhelmed..
- How to detect and correct transmission errors.

Physical factors are considered the interface standard "standard interface"; other factors include "protocols".

The physical method of transferring data across a communication link varies depending on the medium used. The binary values 0 and 1, for example, can be signaled by the presence or absence of voltage on a copper wire, by a pair of sound signals generated and decoded by a modem in the case of the telephone system, or by the use of modulated light in the case of optical fiber [69, 70].

III.3.2 Communication methods:

In any communication scenario involving two devices [71], data can be sent in one of three modes. These are:

III.3.2.1 Simplex :

A simplex system is designed to send messages in only one direction.

III.3.2.2 Half-duplex :

A duplex system is designed to send messages in both directions, but only in one direction at a time.

III.3.2.3 Full-duplex:

In a full-duplex system, data can flow in both directions simultaneously. Figure III.5.

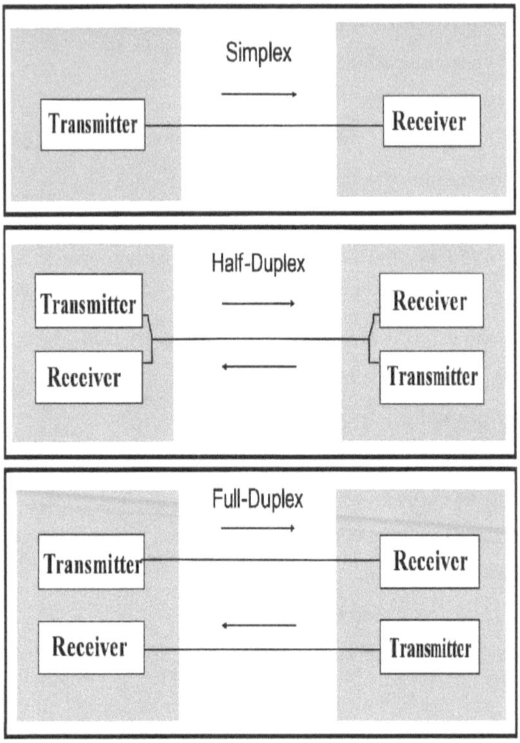

Figure III.5 : Serial communication methods in microcontrollers.

III.3.3 Asynchronous and synchronous communication:

In an asynchronous protocol, the interface lacks a clock line. Instead, each computer has its own clock to use as a synchronization reference.

In contrast, in a synchronous protocol, the interface has a clock line, usually controlled by one of the computers, and all transmitted bits are synchronized to it. Each transmitted bit is valid at a defined time after a rising or falling clock edge, depending on the protocol. Examples of synchronous serial interfaces are I2C, SPI, and Microwire [72, 73].

III.3.4 Word format in the UART module:

A UART transmits data in chunks often called words. Each word contains a start bit, the data bits, an optional parity bit, and one or more stop bits [74].

Most UARTs support multi-word formats. A standard format is 8-N-1, where the transmitter sends each word with a start bit, followed by 8 data bits, and a stop bit. The data transmission bits begin with bit 0 (the least significant bit, or LSB). Figure III.6 illustrates this.

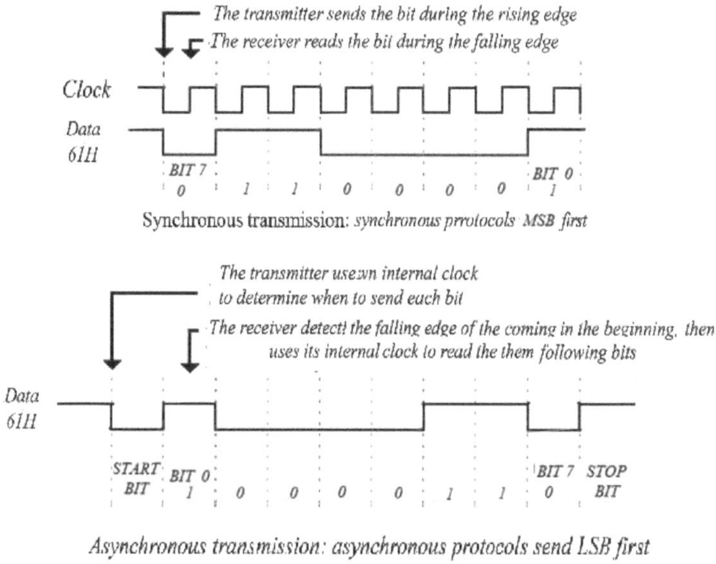

Figure III.6 : Synchronous transmissions include a clock line, whereas asynchronous transmissions require each computer to have its own clock.

The N in 8-N-1 indicates that the words do not contain a parity bit. Parity-bit formats can use even or odd parity. One example of a format using parity is 7-E-1, in which the sender sends a start bit, 7 data bits, an even parity bit, and a stop bit.

The parity bit can provide a basic form of error detection. With equal parity, the data and parity bits in each word contain an even number of 1s. With odd parity, the data and parity bits in each word combine to form an odd number of 1s. For example, suppose the data bits to be sent are 0000001b. With even parity, the parity bit is 1 to make the total number of 1s in the data and parity bits even. With odd parity, the parity bit is 0 to keep the total number of 1s odd. If the data bits are 0000011b, with even parity, the parity bit is 0, and with odd parity, it is 1. If the communications are using 7-E-1 format, a receiving computer that receives a byte with an odd number of 1s knows the data was not transmitted correctly.

III.3.5 The USART or SCI serial link:

The SCI (Serial Communication Interface) link is an asynchronous, START/STOP-type serial interface. Figure III.7 shows that the USART enables communication with other systems or devices over long distances (a few meters to a few hundred meters) [75].

It has the following features:

- Whole Duplex operation, meaning transmission and reception of data at the same time.
- Data transmission and reception (compatible with the RS232 standard using a level adaptation function).
- Error control for transmission and reception.
- Transmission over 8 or 9 bits.
- Automatic wake-up mode upon receipt of valid signals.
- 4 Possible sources of interruptions.

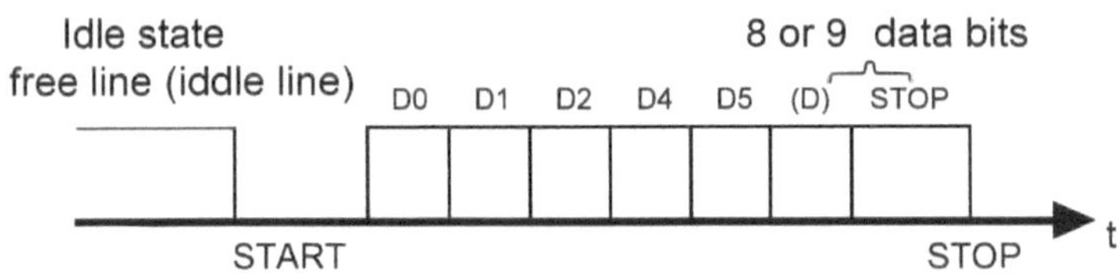

Figure III.7 : The USART module frame in the PIC 16F877.

III.3.5.1 The transmission :

This function uses a shift register to transmit the 8 or 9 bits of information from the TXREG register. For this function (figure III.8) to be operational, the RC6 pin of PORTC must be configured as an output, and the SPEN bit of the RCSTA register and the TXEN bit of the TXSTA register must be set to 1. In the case where a 9-bit transmission is used (Parity bit, for example), the 9-bit transmission must be authorized via the TX9 bit of the TXSTA register. The value of the ninth bit must be recorded in the TX9D bit of the TXSTA register.

Figure III.9 shows that before transmitting information, the transmission register must be free, as indicated by the TXIF bit (1 = free, 0 = occupied). The TMRT bit in the TXSTA register indicates whether the transmission is complete (1 indicates complete, zero indicates occupied). The figure illustrates this.

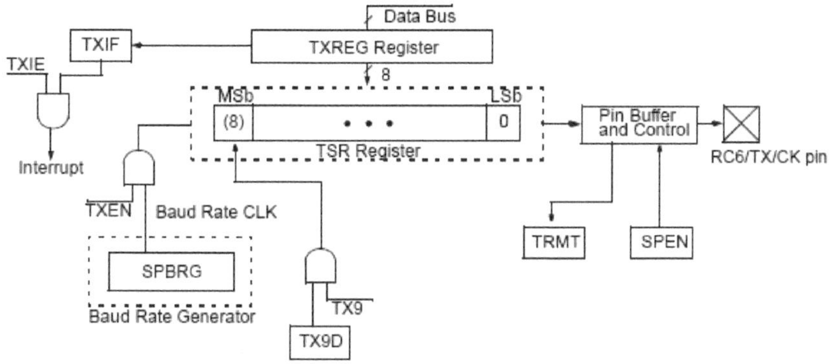

Figure III.8 : USART transmitter block diagram.

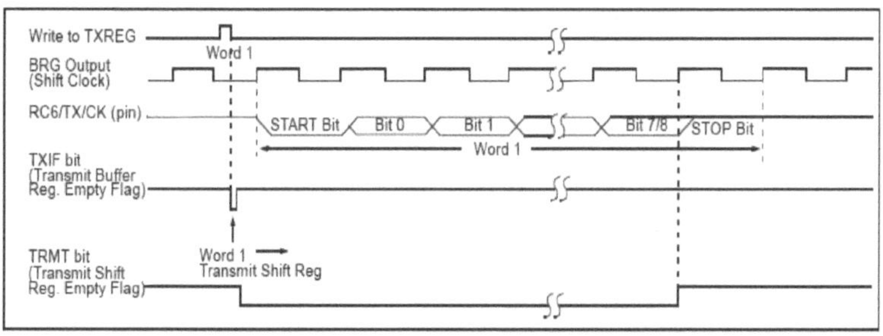

Figure III.9 : The asynchronous transmitter master.

- **The TXSTA register**:

This is the most essential register for transmitting the diagram of this register, which is shown in figure III.10:

TXSTA: TRANSMIT STATUS AND CONTROL REGISTER (ADDRESS 98h)

R/W-0	R/W-0	R/W-0	R/W-0	U-0	R/W-0	R-1	R/W-0
CSRC	TX9	TXEN	SYNC	—	BRGH	TRMT	TX9D
bit 7							bit 0

bit 7 **CSRC:** Clock Source Select bit
Asynchronous mode:
Don't care
Synchronous mode:
1 = Master mode (clock generated internally from BRG)
0 = Slave mode (clock from external source)

bit 6 **TX9**: 9-bit Transmit Enable bit
1 = Selects 9-bit transmission
0 = Selects 8-bit transmission

bit 5 **TXEN**: Transmit Enable bit
1 = Transmit enabled
0 = Transmit disabled

Note: SREN/CREN overrides TXEN in SYNC mode.

bit 4 **SYNC**: USART Mode Select bit
1 = Synchronous mode
0 = Asynchronous mode

bit 3 **Unimplemented:** Read as '0'

bit 2 **BRGH**: High Baud Rate Select bit
Asynchronous mode:
1 = High speed
0 = Low speed
Synchronous mode:
Unused in this mode

bit 1 **TRMT**: Transmit Shift Register Status bit
1 = TSR empty
0 = TSR full

bit 0 **TX9D:** 9th bit of Transmit Data, can be parity bit

Legend:			
R = Readable bit	W = Writable bit	U = Unimplemented bit, read as '0'	
- n = Value at POR	'1' = Bit is set	'0' = Bit is cleared	x = Bit is unknown

Figure III.10 : TXSTA registry configuration.

III.3.5.2 The reception :

This function uses an RSR shift register to receive 8 or 9 bits of information; upon completion of reception, the value is stored in the RCREG register. For this function to be operational, the RC7 pin of PORTC must be configured as an input, and the SPEN bit of the RCSTA register must be set to 1.

When a 9-bit reception is used, it is necessary to enable 9-bit reception via the RX9 bit of the RCSTA register; the value of the ninth bit is retrieved in the RX9D bit of the RCSTA register [76].

Before reading information from the RCREG register, ensure the RCIF bit (1 Data received) is set by testing it; it is reset to 0 when reading the RCREG register.

The FERR and OERR bits can indicate a format error and an "overrun" error, respectively. figures III.11 and III.12 illustrate this.

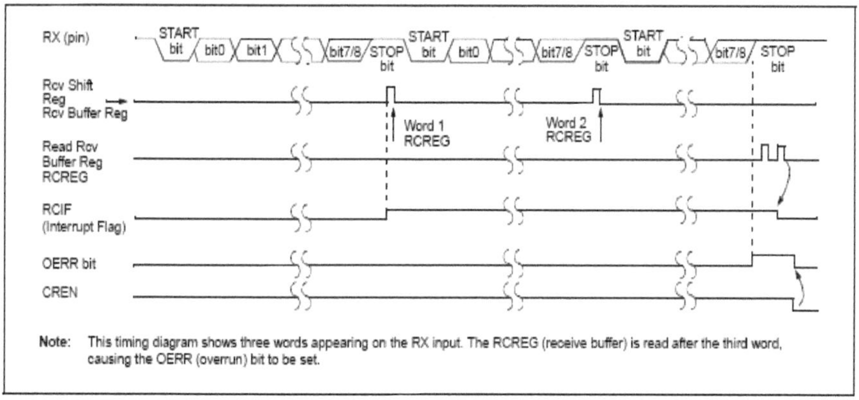

Figure III.11: Asynchronous reception.

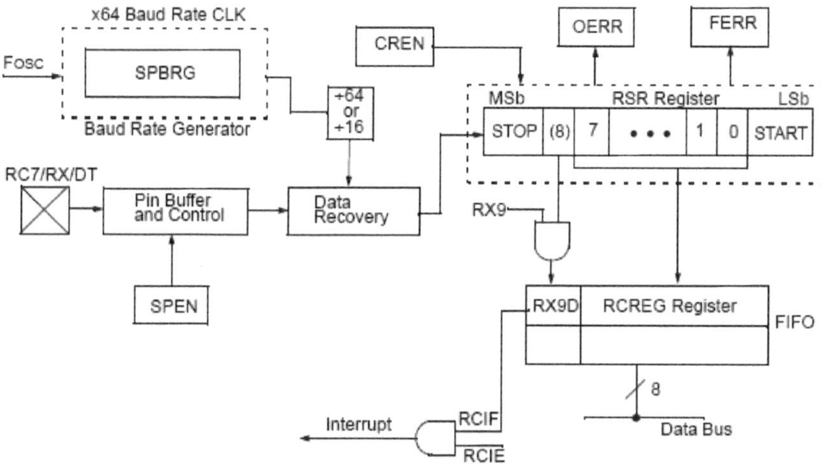

Figure III.12 : The asynchronous receptor.

•The RCSTA register:

The most essential registry for configuring reception is shown in figure III.13:

RCSTA: RECEIVE STATUS AND CONTROL REGISTER (ADDRESS 18h)

R/W-0	R/W-0	R/W-0	R/W-0	R/W-0	R-0	R-0	R-x
SPEN	RX9	SREN	CREN	ADDEN	FERR	OERR	RX9D
bit 7							bit 0

bit 7 **SPEN:** Serial Port Enable bit
1 = Serial port enabled (configures RC7/RX/DT and RC6/TX/CK pins as serial port pins)
0 = Serial port disabled

bit 6 **RX9**: 9-bit Receive Enable bit
1 = Selects 9-bit reception
0 = Selects 8-bit reception

bit 5 **SREN**: Single Receive Enable bit
Asynchronous mode:
Don't care
Synchronous mode - master:
1 = Enables single receive
0 = Disables single receive
This bit is cleared after reception is complete.
Synchronous mode - slave:
Don't care

bit 4 **CREN**: Continuous Receive Enable bit
Asynchronous mode:
1 = Enables continuous receive
0 = Disables continuous receive
Synchronous mode:
1 = Enables continuous receive until enable bit CREN is cleared (CREN overrides SREN)
0 = Disables continuous receive

bit 3 **ADDEN:** Address Detect Enable bit
Asynchronous mode 9-bit (RX9 = 1):
1 = Enables address detection, enables interrupt and load of the receive buffer when RSR<8> is set
0 = Disables address detection, all bytes are received, and ninth bit can be used as parity bit

bit 2 **FERR**: Framing Error bit
1 = Framing error (can be updated by reading RCREG register and receive next valid byte)
0 = No framing error

bit 1 **OERR**: Overrun Error bit
1 = Overrun error (can be cleared by clearing bit CREN)
0 = No overrun error

bit 0 **RX9D:** 9th bit of Received Data (can be parity bit, but must be calculated by user firmware)

Legend:			
R = Readable bit	W = Writable bit	U = Unimplemented bit, read as '0'	
- n = Value at POR	'1' = Bit is set	'0' = Bit is cleared	x = Bit is unknown

Figure III.13 : The RCSTA registry configuration.

III.3.5.3 The clock generator:

It is the one who sets the speed of reception and transmission of the USART (table III.4). In table III.5 and table III.6, the formulas of MODES (BRGH = 0) and (BRGH = 1) must be used to calculate the value to be put in the SPBRG register [77].

SYNC	BRGH = 0 (Low Speed)	BRGH = 1 (High Speed)
0	(Asynchronous) Baud Rate = Fosc/(64(X+1))	Baud Rate = Fosc/(16(X+1))
1	(Synchronous) Baud Rate = Fosc/(4(X+1))	N/A

X = value in SPBRG (0 to 255)

Table III.4: The different cases of the clock generator.

BAUD RATE (K)	Fosc = 20 MHz		
	KBAUD	% ERROR	SPBRG value (decimal)
0.3	-	-	-
1.2	1.221	1.75	255
2.4	2.404	0.17	129
9.6	9.766	1.73	31
19.2	19.531	1.72	15
28.8	31.250	8.51	9
33.6	34.722	3.34	8
57.6	62.500	8.51	4
HIGH	1.221	-	255
LOW	312.500	-	0

Table III.5: Speed table: MODE (BRGH = 0).

BAUD RATE (K)	Fosc = 20 MHz		
	KBAUD	% ERROR	SPBRG value (decimal)
0.3	-	-	-
1.2	-	-	-
2.4	-	-	-
9.6	9.615	0.16	129
19.2	19.231	0.16	64
28.8	29.070	0.94	42
33.6	33.784	0.55	36
57.6	59.524	3.34	20
HIGH	4.883	-	255
LOW	1250.000	-	0

Table III.6 : Speed table: MODE (BRGH = 1).

•The associated registers:

The SCI clock generator configuration is handled through the following registers: TXSTA (address 98h, page 1), RCSTA (address 18h, page 0), and SPBRG (address 99h, page 1).

III.3.6 The RS-232 connection:

Today, the majority of equipment available on the market is based on the RS-232-C or RS-232-D standard. The RS-232C interface standard was published in the United States in 1969 by the Electronics Industries Alliance (EIA) to define the electrical and mechanical details of the interface between data terminal equipment (DTE) and data communication equipment (DCE) for binary data exchange. RS-232, as shown in Figure III.14, is designed to handle communications between two devices with a distance limit of approximately 2438.4 to 3962.4 cm, depending on the bit rate and cable type. Older PCs and modems used a 25-pin male connector. The 9-pin interface first appeared commercially on AT-class PCs in the early 1980s.

The Designation and link in the microcontroller in table III.7 are devised in these parts:

DTE (Data terminal equipment): a data transmission from a terminal such as a computer, which is the source of the data (usually a series of characters encoded in a suitable digital form)

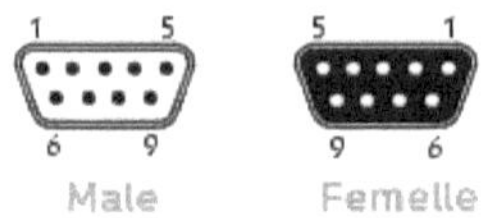

Figure III.14 : DB9 connector.

Number	**Name**	**Designation**
1	CD - Carrier Detect	Carrier detection
2	RXD - Receive Data	Data reception
3	TXD - Transmit Data	Data transmission
4	DTR - Data Terminal Ready	Terminal ready
5	GND - Signal Ground	Logical mass
6	DSR - Data Set Ready	Data ready
7	RTS - Request To Send	Issuance request
8	CTS - Clear To Send	Ready to issue
9	RI - Ring Indicator	Ring indicator

Table III.7 : The roles of pins.

DCE (Data Communications Equipment or Data Carrier Equipment): which acts as a data converter (for example, a modem; in our case, a PIC) to convert the signal into a form suitable for the communication link, as shown in figure III.15, for example, analog signals for the telephone system. There are three essential lines for bidirectional RS-232 communications:

TX: transports data from the DTE to the DCE, sometimes referred to as TD or TXD.

RX: transports data from the DCE to the DTE, sometimes named RD or RXD.

SG: The ground signal, sometimes called GND or SGND.

The use of the signal remainders indicated in Table III.7 depends on the application we want to manipulate [78].

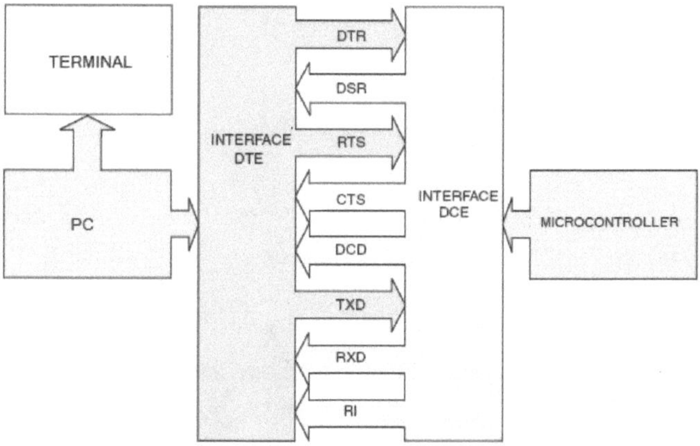

Figure III.15 : PC, RS232, Microcontroller interface.

III.4 Conclusion :

The conversion method used by our module is successive approximation with a serial output. This operation uses the binary search method to convert an analog voltage into a digital format, offering greater efficiency than exhaustive search of the ramp counter and fast operation, but this speed introduces architectural complexity.

The advantage of serial communication is that it requires simpler transmission methods than parallel communication, which transmits the same data. Parallel communication requires the ability to communicate as many bits as there are pieces of information, while serial communication can handle any number of bits (even just one). However, several problems arise with the parallel transmission of large numbers of bits, such as crosstalk and desynchronization. This serial communication will be implemented by the PIC16F877 (USART), which is specifically designed to communicate using the UART interface.

Chapter 4

Practical Implementation and Signal Acquisition of the ECG System

IV.1 Introduction :

To measure heart rate, our electrocardiogram (ECG) uses a signal captured by three electrodes, with one serving as the reference (ground). The ECG electrical signal originates from the heart's muscular activity. By performing mechanical work through ion exchange (Chapter 1) across its cell wall, each cell becomes an electrical dipole. The elementary electrical fields generated by each heart cell are summed to form the cardiac electrical field. These voltages are typically recorded and then applied to an amplifier before processing [30].

IV.2 Practical diagram of the ECG:

The ECG system is shown in figure IV.1. The ECG system comprises four stages, each stage being as follows:

(1) The first stage is a transducer—AgCl electrode, which converts the ECG into an electrical voltage. (1) he first step is a 1-AgCl transducer electrode, which converts the ECG into an electrical voltage. The voltage is in the range of 1 mV ~ 5 mV.

(2) The second stage is an instrumentation amplifier (Analog Device, AD624), which has a very high CMRR (90dB) and high gain (1000), with a power supply of +9V. (2) The second stage is an instrumentation amplifier (Analog Device, AD624), which has a very high CMRR (90dB) and a high gain (1000), with a +9V power supply.

(3) We use an opto-coupler (NEC PS2506) to isolate the In-Amp and output. (3) We use an optocoupler (NEC PS2506) for isolation of the common-mode output voltage of AMP.

(4) After the optocoupler is a bandpass filter from 0.04Hz to 150Hz. A cascade of low-pass and high-pass filters implements it.

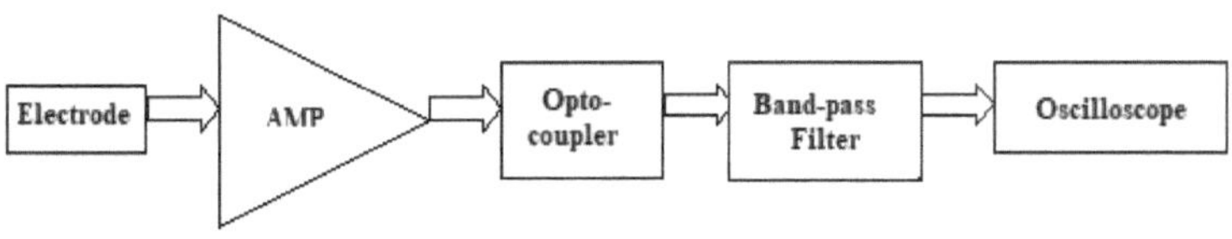

Figure IV.1 : The function blocks of the ECG system.

The ECG is converted into electrical voltage by electrodes. The electrode used for ECG recording is Ag/AgCl. Disposable electrodes are attached to the patient's skin and can be easily removed. The cardiac mechanism of the ECG accounts for the fact that the electrocardiogram (ECG) initiates the cardiac cycle. Heart sounds are shown in figure IV.2 and illustrate that ejection occurs only when the pressure in the left ventricle exceeds that in the aorta [32, 79].

Once the ECG electrodes have converted into electrical voltage, these voltages can be fed into an instrumentation amplifier and then processed.

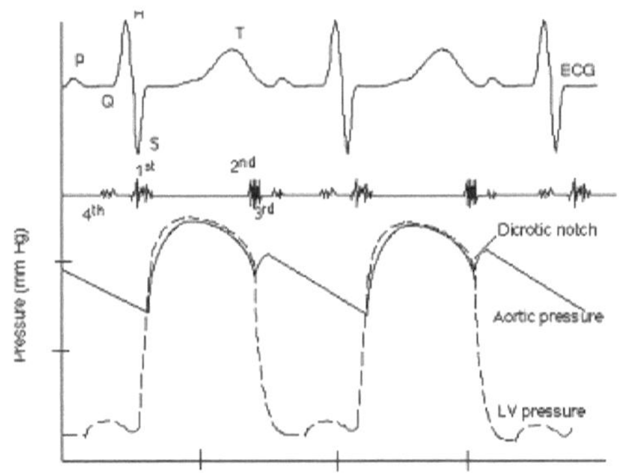

Figure IV.2 : The ECG cardiac cycle.

We measure the ECG by connecting two electrodes to the left and right chests, respectively, as shown in figure IV.3. The body must be grounded, so we connect the leg to the ground. If the body is not grounded, no signal will be obtained.

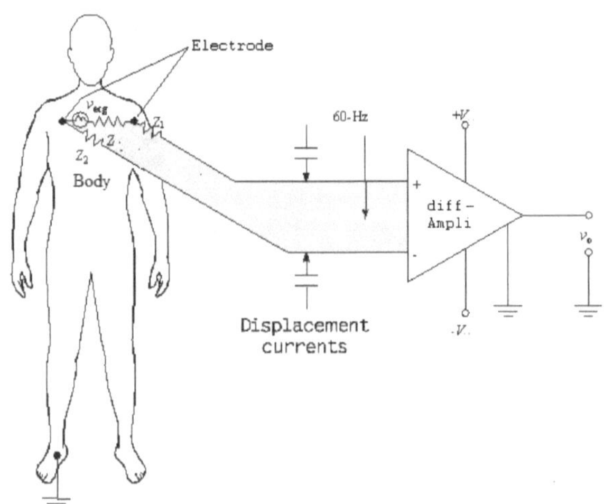

Figure IV.3 : ECG simplified recording system.

IV.3 ECG system circuits:

The ECG circuit diagram is shown in figure IV.4:

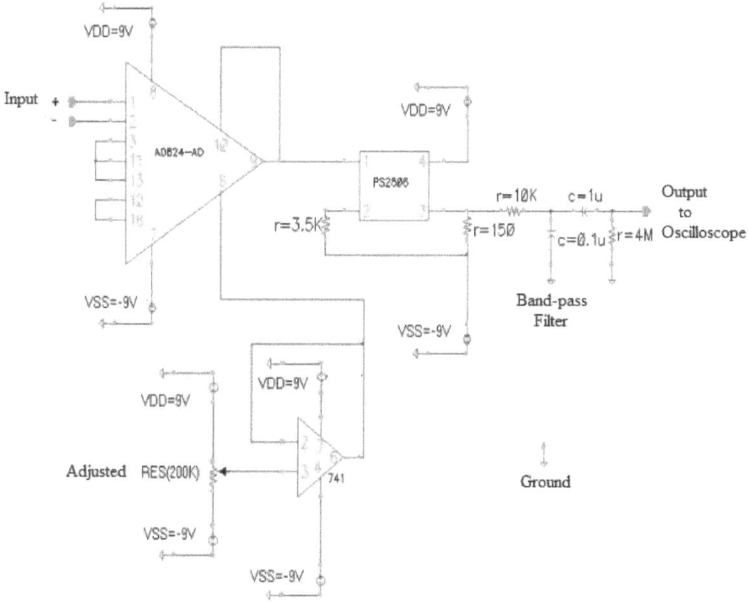

Figure IV.4 : The ECG circuit diagram.

IV.3.1 Instrumentation amplifier:

We chose the Analog Devices AD624 instrumentation amplifier to amplify the ECG electrode voltage, which is on the order of several mV. We decided to use the Analog Devices AD624 instrumentation amplifier to amplify the ECG voltage from the electrodes, which is in the range of several mV. The AD624 is set to a gain of 1000 and is supplied by a +9V battery. Some essential features of the AD624 are listed in Table IV.1 [80].

Supply voltage	**-9V power battery +**
Programmable Gain	1, 100, 200, 500, 1000, 2500
CMRR	130dB (Gain = 500 to 1000)
Product bandwidth gain	25 MHz
Input offset	25 mV, max

Table IV.1 : Key features of the AD624 Instrumentation Amplifier.

A high CMRR is crucial for amplifier instrumentation. The small AC voltage signal (less than five mV) detected by the sensor on the electrodes will be accompanied by a significant common AC mode element (up to 1.5 V) and a significant DC component (300 mV). The common-mode rejection specified by the AAMI (Association for the Advancement of Medical Instrumentation) is 89 dB for standard ECGs and 60 dB for ambulatory recorders. The CMRR of the AD624 with a gain of 1000 is shown in figure IV.5. The CMRR equation:

Gain differential = CMRR / common-mode gain = Adm/ACM

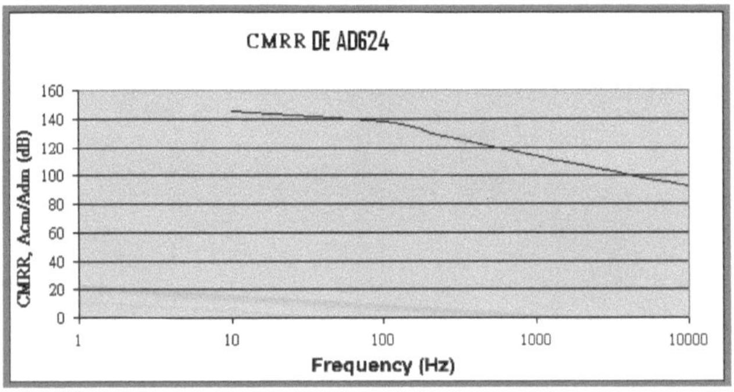

Figure IV.5 : Measured CMRR data AD624 En-Amp.

A 741 op-amp is connected to the reference node (node 6) of the AD624 to compensate for the 3V DC output voltage. The 741 is connected as a source follower. The DC output voltage of the AD624 is set to 3V, an optimal value determined through trial and error.

IV.3.2 Opto-coupler:

After the En-Amp, an NEC optocoupler PS2506 cascades to isolate the En-Amp from the rest of the circuits, as shown in figure IV.4. The datasheet for the NEC PS2506 is on

IV.3.3 Band-pass filter:

In general, components of the signal of interest will reside in the 0.67 to 40-Hz bandwidth for standard ECGs and up to 300 Hz to 1 kHz for pacemaker detection. In general, the signal components of interest will reside within the 40 Hz bandwidth of 0.67 Hz for standard ECG and up to 300 Hz to 1 kHz for pacemaker detection. We adopt the suggestion in John Webster's book to use a bandpass filter in the frequency range of 0.04 Hz to 150 Hz. A cascade of low-pass and high-pass filters implements the filter. The low-pass and high-pass filter data are implemented by simple RC components, as shown in Table IV.2. The measured transfer function of the bandpass filter is shown in Figure IV.6. [81]

	R	C	Time constant T	3dB Frequency
Low-pass filter	4 M Ω	1 µ F	4,0 sec	0,04 Hz
High-pass filter	10 KΩ	0,1 µ F	0.001 sec	159Hz

Table IV.2 : Filter the data.

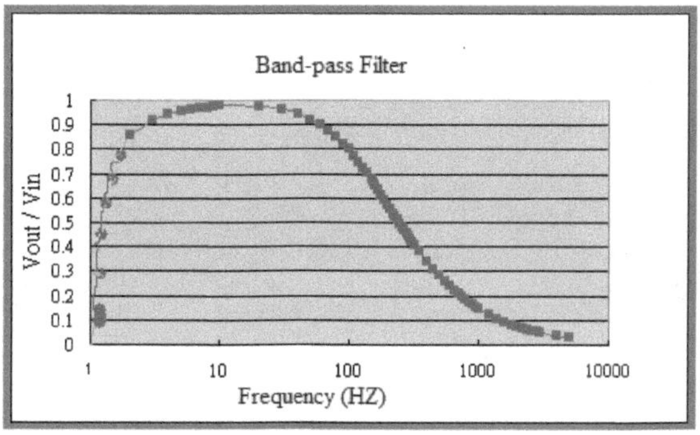

Figure IV.6 : The measured transfer function of the bandpass filter.

IV.4 Measured ECG signals:

The ECG circuit is shown in figure IV.7. We use the Agilent 54624 Oscilloscope to probe the ECG signal. The three ECG signals shown in figure IV.8 are from the instrumentation amplifier, the optocoupler, and the filter, respectively. The circuits function perfectly, as the three signals are nearly identical.

Measuring different regions of the heart will obtain different ECG signals. Measuring different areas of the heart yields different ECG signals. The ECG signal shown in figure IV.7 is standard [82, 83].

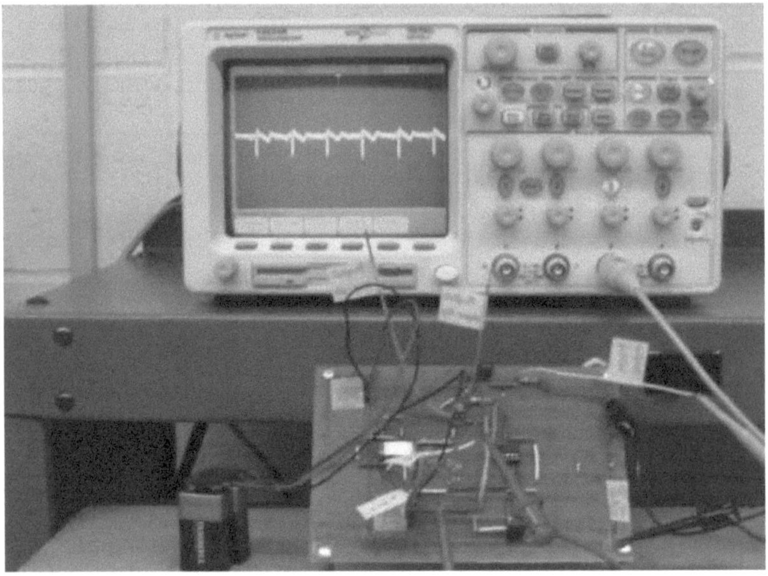

Figure IV.7 : ECG circuits.

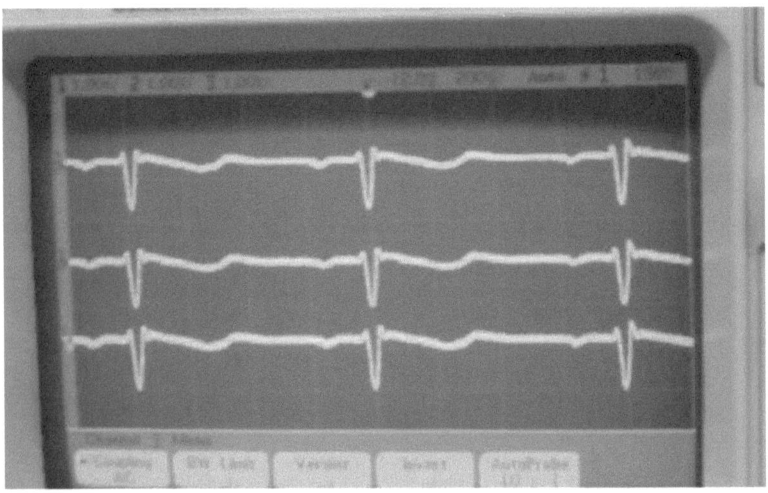

Figure IV.8 : ECG signals. The three signals are respectively probes (1) En-Amp (2) opto-coupler (3) Filter.

IV.5 Signal conversion:

The 16F87x circuit has an analog-to-digital converter modulator that allows 10-bit sampling. The digital signal can therefore take 1024 possible values, and the maximum input voltage of this ADC is 5V. To calculate this, we need to ensure we have all the necessary information. the 10-bit conversion accuracy: 5V/(2^10 - 1) = 4,88785 mV/pas. With an acquisition time: $T_{acq} = T_{amp} + T_c + T_{coff} = 2\mu s + 16{,}46\mu s + 1{,}25\mu s = 19{,}71\mu s$

With: T_{amp} = circuit reaction time, T_c = capacitor charging time, and T_{coff} = time which depends on the temperature coefficient.

The temperature coefficient is typically 0.05 µs per °C above 25°C; this time Tcoff will be at least 0 (below 25°C) and at most (50-25)*0.05, or 1.25 µs. The PIC® temperature cannot, in fact, exceed 50°C.

As shown in figure IV.9, and after the analog input is selected, the modulator performs the conversion procedures, and the result is stored in a digital code format in the ADRESL and ADRESH registers [84, 85]. The oscillation used in our circuit is 20 MHz.

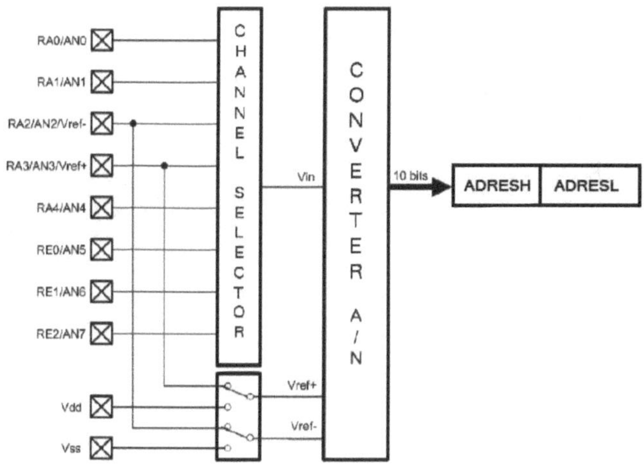

Figure IV.9 : A-D Conversion Modulator.

IV.6 Signal transmission with the USART module in asynchronous mode :

As explained in the previous chapter, an asynchronous serial connection means transmitting or receiving data without a clock signal to synchronize the transmitter and receiver; therefore, we need to know the transmitter's transfer rate and synchronize the receiver to it. Next, we need to establish a mechanism for detecting the beginning of a data transmission.. Alternatively, an excitation against the resting state, represented by the start bit, is required. This is followed by a data stream of 8 or 9 bits (the 9th can be used as a parity bit, either even or odd) [86, 87]. Finally, a mechanism, represented by one or more stop bits, must be established to return the line to its resting state at the end of each transmission. At the PIC level, the connection will use the two pins RC6 (TX/CK) and RC7 (RX/DT). The intermediate circuit MAX232, which is connected to the circuit that generates the ECG signal and our USART module converter, has the role of converting the signals using its internal capacitors (figure IV.10).

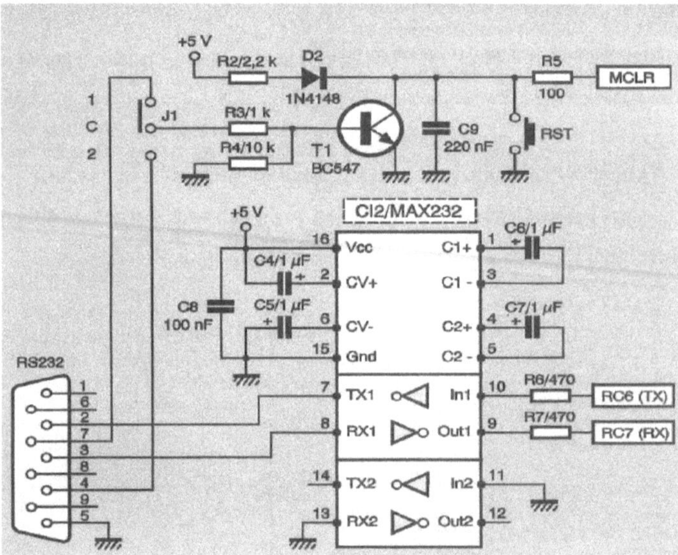

Figure IV.10 : The signal interface section using the Max 232 and RS232 link (development board).

IV.7 Receiving data with the UART module:

On the PC, there is a DB9 connector for the RS232 serial link. Using this connector, we can communicate with the motherboard, which includes a UART modulator [57, 75, 88]. Using Visual Basic .NET software (figure IV.11), we will select the port to use and then open the ports for this module. Storing the received bytes in the buffer (register) is crucial, and it interprets each ten-bit data point on an XY axis. Do not forget the baud rate (reception speed), the type of parity used, and the number of stop bits. The output signal is shown in Figure IV.12. The points on this axis represent the graph of the signal converted and transmitted by our development board. Figure IV.13.

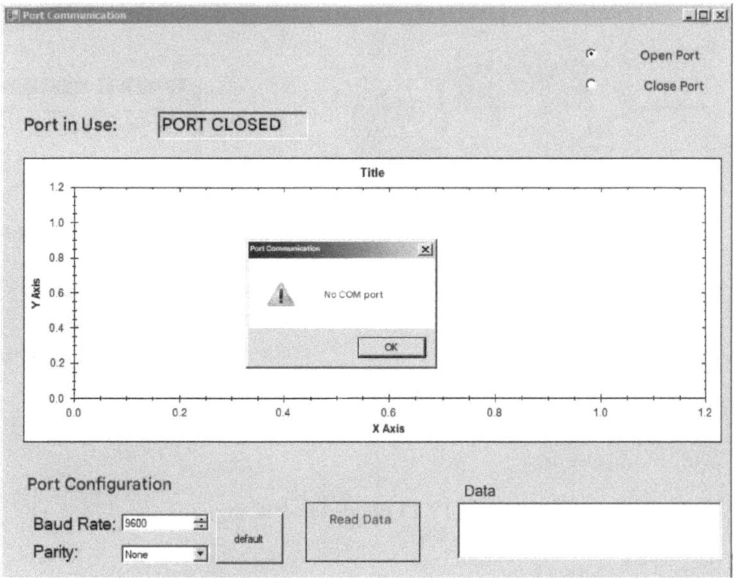

Figure IV.11 : The Visual Basic ports .NET.

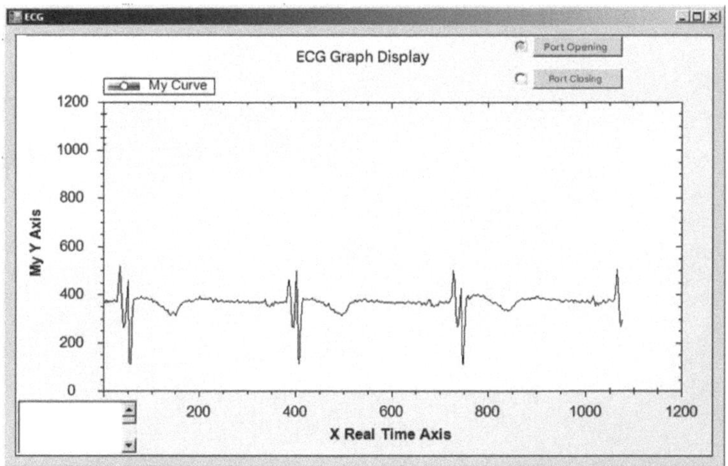

Figure IV.12 : ECG signal plotted.

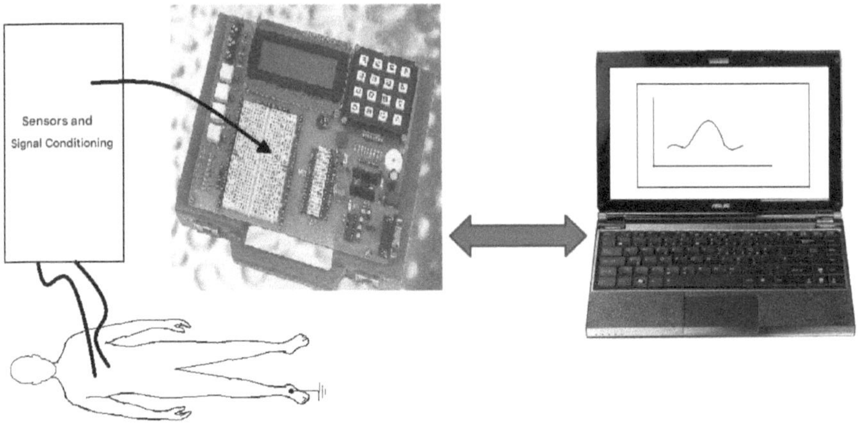

Figure IV.13 : Signal interfacing.

Conclusion and Perspectives

Conclusion and Perspectives :

This work presents the design and implementation of a microcontroller-based electrocardiogram (ECG) acquisition system, situated at the intersection of biomedical engineering and embedded electronics. The project responds to a growing clinical and technological need: the development of portable, low-cost, and reliable tools for cardiac monitoring, particularly in environments where access to advanced diagnostic infrastructure is limited or intermittent.

The heart, as a vital organ, generates bioelectrical signals that reflect its physiological state. Capturing and interpreting these signals through electrocardiography has become a cornerstone of modern cardiology. However, traditional ECG systems are often bulky, expensive, and dependent on hospital-based infrastructure. This project addresses these limitations by proposing a compact, battery-powered ECG device that acquires, amplifies, digitizes, and wirelessly transmits cardiac signals to a computer for visualization and analysis.

The system is built around the **PIC16F877 microcontroller**, chosen for its versatility, low power consumption, and integrated analog-to-digital conversion capabilities. The device architecture includes electrodes for signal acquisition, a differential amplifier for noise reduction, an ADC module for digitization, and a UART interface for serial communication. The signal is then displayed on a PC, allowing clinicians to monitor cardiac activity in real time with enhanced clarity and precision.

From a technical standpoint, the project demonstrates the feasibility of integrating multiple subsystems—electrical, computational, and communicative—into a cohesive and functional biomedical device. The use of **CMOS logic circuits** ensures minimal power consumption and safe operation, particularly important for wearable applications. The system is designed to operate autonomously on a 9V battery, reinforcing its portability and independence from mains power sources, which is crucial for patient mobility and field deployment.

On the software side, the project employs signal processing techniques to improve the quality of acquired ECG data. Filtering algorithms remove noise and artifacts, while waveform detection methods identify key cardiac events, such as QRS complexes, P waves, and T waves. These features enhance the system's diagnostic value and lay the groundwork for future integration of automated analysis and alert mechanisms.

Clinically, the device is intended to support patients experiencing cardiac discomfort or arrhythmias by providing continuous monitoring and early warning of abnormal exertion. By enabling real-time visualization of cardiac signals, the system empowers both patients and healthcare providers to make informed decisions and avoid potentially harmful overexertion. Its portability and simplicity of use make it suitable for deployment in hospitals, outpatient clinics, and even home-based care settings.

Beyond its immediate application, this project opens avenues for further development and innovation. Future iterations could incorporate wireless communication protocols such as Bluetooth or Wi-Fi, cloud-based data storage, and mobile application interfaces. Additionally, Machine learning algorithms can be integrated to provide predictive diagnostics and personalized cardiac profiling.

In conclusion, this modest yet impactful work illustrates how embedded systems can be harnessed to address real-world medical challenges. It bridges the gap between theoretical knowledge and practical implementation, demonstrating that with thoughtful design and interdisciplinary collaboration, it is possible to create tools that enhance patient care, expand diagnostic accessibility, and contribute meaningfully to the evolution of biomedical technology.

References

References:

[1] Ren, M. X., Zeng, Y., Nowlen, P., & Zhou, Q. (2025). Advancements in Flexible and Wearable Echocardiograms for Real-Time Continuous Cardiovascular Monitoring. Current Treatment Options in Cardiovascular Medicine, 27(1), 1-12.

[2] Pandey, G., & Vora, A. (2019). Open electronics for medical devices: state-of-art and unique advantages. Electronics, 8(11), 1256.

[3] Nandini, K. P., & Seshikala, G. (2021, November). Role of embedded computing systems in biomedical applications–opportunities and challenges. In 2021 IEEE International Conference on Distributed Computing, VLSI, Electrical Circuits and Robotics (DISCOVER) (pp. 332-335). IEEE.

[4] Zhao, H. (2023). Applications of embedded systems in medicine: Challenges and future trends. Highlights in Science, Engineering and Technology, 62, 31-35.

[5] Banchs, J. E., & Scher, D. L. (2015). Emerging role of digital technology and remote monitoring in the care of cardiac patients. Medical Clinics, 99(4), 877-896.

[6] Deopujari, S. P., & Gaikwad, A. (2013). Light weight, low cost, wearable ECG monitoring. Int J Comput Eng Res, 3(7), 1-6.

[7] Sarkar, D., & Chowdhury, A. (2015, January). Low cost and efficient ECG measurement system using PIC18F4550 microcontroller. In 2015 International Conference on Electronic Design, Computer Networks & Automated Verification (EDCAV) (pp. 6-11). IEEE.

[8] Qureshi, R., Nawaz, M., Khuhawar, F. Y., Tunio, N., & Uzair, M. (2019). Analysis of ECG signal processing and filtering algorithms. International Journal of Advanced Computer Science and Applications, 10(3), 545-550.

[9] Lee, E. K., Wang, Y., Davis, R. A., & Egan, B. M. (2017, November). Designing a low-cost adaptable and personalized remote patient monitoring system. In 2017 IEEE International Conference on Bioinformatics and Biomedicine (BIBM) (pp. 1040-1046). IEEE.

[10] Khairurrijal, Abdullah, M., & Budiman, M. (2011). Home-made PIC 16F877 microcontroller-based temperature control system for learning automatic control. Computer Applications in Engineering Education, 19(1), 10-17.

[11] Hampton, J., Hampton, J., & Adlam, D. (2024). The ecg made easy e-book: The ecg made easy e-book. Elsevier Health Sciences.

[12] Wagner, G. S., & Strauss, D. G. (2013). Marriott's practical electrocardiography. Lippincott Williams & Wilkins.

[13] Surawicz, B., & Knilans, T. (2008). Chou's electrocardiography in clinical practice: adult and pediatric. Elsevier Health Sciences.

[14] Guyton, A. C. (2021). Hall JE Textbook of Medical Physiology.–14-nashr.

[15] Netter, F. H. (2018). Atlas of Human Anatomy: Latin Terminology E-Book: Atlas of Human Anatomy: Latin Terminology E-Book. Elsevier Health Sciences.

[16] Drake, R. L., Vogl, A. W., & Mitchell, A. W. (2019). Gray's Anatomy for Students Flash Cards E-Book: Gray's Anatomy for Students Flash Cards E-Book. Elsevier Health Sciences.

[17] Zipes, D. P., Jalife, J., & Stevenson, W. G. (2017). Cardiac electrophysiology: from cell to bedside E-book. Elsevier Health Sciences.

[18] Beheshti, M., Umapathy, K., & Krishnan, S. (2016). Electrophysiological cardiac modeling: a review. Critical Reviews™ in Biomedical Engineering, 44(1-2).

[19] Anderson, R. H., Yanni, J., Boyett, M. R., Chandler, N. J., & Dobrzynski, H. (2009). The anatomy of the cardiac conduction system. Clinical Anatomy: The Official Journal of the American Association of Clinical Anatomists and the British Association of Clinical Anatomists, 22(1), 99-113.

[20] Bers, D. (2012). Excitation-contraction coupling and cardiac contractile force (Vol. 237). Springer Science & Business Media.

[21] Hille, B. (2001). Ion channels of excitable membranes third edition. *(No Title)*.

[22] Clancy, C. E., & Rudy, Y. (2001). Cellular consequences of HERG mutations in the long QT syndrome: precursors to sudden cardiac death. *Cardiovascular research*, *50*(2), 301-313.
[23] Ravens, U., Gomez, A. M., Heijman, J., Remme, C. A., Dobrev, D., Smith, G., ... & Sipido, K. R. (2021). Edward Carmeliet (1930–2021)—channelling scientific curiosity: a tribute from the ESC Working Group on Cardiac Cellular Electrophysiology.
[24] Fisch, C. (2000). Centennial of the string galvanometer and the electrocardiogram. Journal of the American College of Cardiology, 36(6), 1737-1745.
[25] Kuijpers, P. M. J. C. (2021). History in medicine: the road to clinical electrophysiology. E-Journal of Cardiology Practice December, 21(13–15).
[26] Goldberger, A. L., Goldberger, Z. D., & Shvilkin, A. (2023). *Goldberger's Clinical Electrocardiography-E-Book: A Simplified Approach*. Elsevier Health Sciences.
[27] Kusumoto, F. (2020). ECG interpretation: from pathophysiology to clinical application. Springer Nature.
[28] Wagner, G. S., Macfarlane, P., Wellens, H., Josephson, M., Gorgels, A., Mirvis, D. M., ... & Gettes, L. S. (2009). AHA/ACCF/HRS recommendations for the standardization and interpretation of the electrocardiogram: part VI: acute ischemia/infarction a scientific statement from the American Heart Association Electrocardiography and Arrhythmias Committee, Council on Clinical Cardiology; the American College of Cardiology Foundation; and the Heart Rhythm Society Endorsed by the International Society for Computerized Electrocardiology. Journal of the American College of Cardiology, 53(11), 1003-1011.
[29] Bayés de Luna, A. (2012). Clinical electrocardiography: a textbook.
[30] Webster, J. G. (Ed.). (2009). *Medical instrumentation: application and design*. John Wiley & Sons.
[31] Sansen, W. (2014, February). SC1: Biomedical and sensor interface circuits. In *2014 IEEE International Solid-State Circuits Conference Digest of Technical Papers (ISSCC)* (pp. 528-528). IEEE.
[32] Kalevo, L., Miettinen, T., Leino, A., Kainulainen, S., Myllymaa, K., Töyräs, J., ... & Myllymaa, S. (2019). Improved sweat artifact tolerance of screen-printed EEG electrodes by material selection-comparison of electrochemical properties in artificial sweat. *Ieee Access*, *7*, 133237-133247.
[33] Nayak, S. K., Banerjee, I., & Pal, K. (2019). Electrocardiogram signal processing-based diagnostics: applications of wavelet transform. In Bioelectronics and Medical Devices (pp. 591-614). Woodhead Publishing.
[34] Pan, J., & Tompkins, W. J. (2007). A real-time QRS detection algorithm. IEEE transactions on biomedical engineering, (3), 230-236.
[35] Mckinlay, R., Causey, D., & Mazidi, M. A. (2008). PIC Microcontroller and Embedded Systems: Using assembly and C for PIC 18.
[36] Peatman, J. B. (1998). *Design with PIC microcontrollers*. Pearson Education India.
[37] Stallings, W. (2021). *Computer Organization and Architecture, EBook [Global Edition]*. Pearson Education, Limited.
[38] Patterson, D. A., & Hennessy, J. L. (2016). *Computer organization and design ARM edition: the hardware software interface*. Morgan kaufmann.
[39] Bello, R. W., & Olubummo, D. A. (2019). Simplified traffic lights using 8051 maxim DS89C4XX embedded controller (MDE). *Asian Journal of Mathematical Sciences*.
[40] Ganssle, J. (2008). *The art of designing embedded systems*. Newnes.
[41] KOCER, S., & BUTUNER, R. (2021). Pic Microcontroller. *Programmable Smart Microcontroller Cards*, 126.
[42] Wilmshurst, T. (2006). *Designing embedded systems with PIC microcontrollers: principles and applications*. Elsevier.
[43] Raghunathan, K. R. (2021). History of microcontrollers: First 50 years. *IEEE Micro*, *41*(6), 97-104.
[44] Dhamansure, K., & Desai, S. G. (2020). A Framework for Developing a Microchip PIC Microcontroller Based Application.
[45] Pedram, M., Piguet, C., De, V., Verbauwhede, I., Miranda, M., Moshnyaga, V., ... & Macii, A. (2018). Low-Power Electronics Design.

[46] Horowitz, P., Hill, W., & Robinson, I. (1989). *The art of electronics* (Vol. 2, p. 658). Cambridge: Cambridge university press.
[47] Sedra, A., Smith, K. C., Carusone, T. C., & Gaudet, V. (2020). Microelectronic circuits 8th edition. *Chapter*, *14*, 1235-1236.
[48] Woo-Garcia, R. M., Argüelles-Lucho, P., de Oca-Mora, N. J. M., Salas-Rodriguez, S., Sanchez-Vidal, A., Ceron-Alvarez, C. A., & Osorio-de la Rosa, E. (2024, February). Evaluation of assembler and c programming languages on pic16f877 microcontroller. In *Journal of Physics: Conference Series* (Vol. 2699, No. 1, p. 012013). IOP Publishing.
[49] Alammari, K., Ahmadi, A., & Ahmadi, M. (2020, November). Hybrid memristor-CMOS based up-down counter design. In *2020 27th IEEE International Conference on Electronics, Circuits and Systems (ICECS)* (pp. 1-4). IEEE.
[50] Andrei, M., Belea, R., & Nistor, N. (2019, October). Optimization of a Domestic Microcontroller Application that Operates with Long Standby Time Intervals. In *2019 IEEE 25th International Symposium for Design and Technology in Electronic Packaging (SIITME)* (pp. 291-294). IEEE.
[51] Sladká, K., Czebe, J., & Wagnerová, R. (2018, May). The 2DOF controller implementation in control system with single-chip computer. In *2018 19th International Carpathian Control Conference (ICCC)* (pp. 228-234). IEEE.
[52] Ibrahim, D. (2011). *Advanced PIC microcontroller projects in C: from USB to RTOS with the PIC 18F Series*. Newnes.
[53] Morshed, B. I. (2021). Hardware Design. In *Embedded Systems–A Hardware-Software Co-Design Approach: Unleash the Power of Arduino!* (pp. 37-100). Cham: Springer International Publishing.
[54] Sklar, B. (2021). Digital communications: fundamentals and applications. Pearson.
[55] Proakis, J. G., & Salehi, M. (2001). Digital communications (Vol. 4, pp. 593-620). New York: McGraw-hill.
[56] Haykin, S. (2008). Communication systems. John Wiley & Sons.
[57] Türe, K., Dehollain, C., & Maloberti, F. (2020). Wireless Power Transfer and Data Communication for Intracranial Neural Recording Applications. Springer Nature.
[58] Schreier, R., & Temes, G. C. (2005). Understanding delta-sigma data converters (Vol. 74). Piscataway, NJ: IEEE press.
[59] Carusone, T. C., Johns, D. A., & Martin, K. W. (2011). *Analog integrated circuit design*. John Wiley & Sons.
[60] Baker, R. J. (2019). CMOS: circuit design, layout, and simulation. John Wiley & Sons.
[61] Allen, P. E., & Holberg, D. R. (2011). *CMOS analog circuit design*. Elsevier.
[62] Ghassoul, M. (2020). Design of Three-Term Controller Using a PIC18F452 Microcontroller. In *Renewable Energy-Technologies and Applications*. IntechOpen.
[63] Di Jasio, L., Wilmshurst, T., Ibrahim, D., Morton, J., Bates, M. P., Smith, J., ... & Hellebuyck, C. (2007). PIC Microcontrollers: know it all. Newnes.
[64] Bates, M. P. (2008). *Programming 8-bit PIC microcontrollers in C: with interactive hardware simulation*. Newnes.
[65] Mode, S., Generation, M. D., & Audit, I. I. IIND SEMESTER. *GIET UNIVERSITY*, 22.
[66] Van de Plassche, R. J. (2013). *CMOS integrated analog-to-digital and digital-to-analog converters* (Vol. 742). Springer Science & Business Media.
[67] Candy, J. C., & Temes, G. C. (Eds.). (1991). *Oversampling delta-sigma data converters: theory, design, and simulation*. John Wiley & Sons.
[68] Maheshwari, S. (2023). *Analog Circuit Design Using Current-Mode Techniques*. CRC Press.
[69] Couch, L. W., Kulkarni, M., & Acharya, U. S. (2013). Digital and analog communication systems (Vol. 8). Upper Saddle River: Pearson.
[70] Babajans, R., Cirjulina, D., Capligins, F., Kolosovs, D., Grizans, J., & Litvinenko, A. (2023). Performance analysis of vilnius chaos oscillator-based digital data transmission systems for iot. *Electronics*, *12*(3), 709.
[71] Varsha, P. S., & Hari, V. S. (2022, November). LSTM based receiver design for baseband signal demodulation. In *2022 International Conference on Computing, Communication, and Intelligent Systems (ICCCIS)* (pp. 713-718). IEEE.

[72] Freeman, L. (2020). Test and evaluation for artificial intelligence. *Insight*, *23*(1), 27-30.
[73] Lathi, B. P., & Ding, Z. (2019). *Modern digital and analog communication systems*. Oxford university press.
[74] Martin, J. N. (2020). *Systems engineering guidebook: A process for developing systems and products*. CRC press.
[75] Axelson, J. (2007). *Serial port complete: COM ports, USB virtual COM ports, and ports for embedded systems*. Lakeview Research.
[76] Part, S. N., & Hrs, I. B. Sc., ELECTRONICS SCIENCE.
[77] Tasci, B. (2024). A Survey: Internet of Things (IoTs) Technologies, Embedded Systems and Sensors. In *Communication Technologies and Security Challenges in IoT: Present and Future* (pp. 173-195). Singapore: Springer Nature Singapore.
[78] Chen, D. (2024, October). Research and Design of a Robot Controller Based on MicroPython. In *2024 17th International Congress on Image and Signal Processing, BioMedical Engineering and Informatics (CISP-BMEI)* (pp. 1-6). IEEE.
[79] Reddy, D. C. (2005). Biomedical signal processing: principles and techniques.
[80] Bronzino, J. D. (2000). *Biomedical engineering handbook 2* (Vol. 2). Springer Science & Business Media.
[81] Rubtsov, Y. V., & Zhmud, V. A. (2025). Review Not peer-reviewed version.
[82] Deliyannis, T., Sun, Y., & Fidler, J. K. (2019). *Continuous-time active filter design*. CRC press.
[83] Rahbar, A., Abasi, F., Niksirat, P., Hosseinpour, S., & Seyfollahi, S. (2010). Biomedical Monitoring Systems [heart monitoring system].
[84] Litovski, V. (2019). *Electronic filters*. Springer Singapore.
[85] Dimopoulos, H. G. (2011). *Analog electronic filters: theory, design and synthesis*. Springer Science & Business Media.
[86] Kotov, M. A., Kozlov, P. V., Gerasimov, G. Y., Levashov, V. Y., Shemyakin, A. N., Solovyov, N. G., ... & Malyutin, A. M. (2023). Thermoelectric detector application for measuring the ignition delay time in a shock heated combustible mixture. *Acta Astronautica*, *204*, 787-793.
[87] Qureshi, W. A. (2020). An Extended-Range Hybrid Analog-to-Digital Converter for Audio Applications.
[88] Gözütok, M. E. (2010). Designing programmable logic controller for data acquisition and control (Master's thesis, Izmir Institute of Technology (Turkey)).

Appendix

Appendix

• GLOSSARY

Electrogenesis: the electrical properties of the cell

Cardiovascular: the heart's circulatory system

Lead: an electrical circuit defined by a pair of electrodes placed at various points on the body

EKG: common abbreviation for electrocardiogram, from the German Elektrokardiografie

Stress electrocardiogram: The electrocardiograph is connected to a stationary bike or a treadmill. This way it knows what power the patient is exerting at each moment, and can thus relate heartbeats to the effort provided. Programs are set up, and the device controls the treadmill's incline or the stationary bike's resistance.

String galvanometer: a highly sensitive mechanical ammeter that has a needle indicating the direction and strength of the current. Recordings are made with a stylus fixed to the needle.

Thermal printing: the printer runs a heated nozzle over special paper (called thermal paper, which is heat-sensitive). The nozzle heats very specific points in a few thousandths of a second. Traces are preserved for around ten years if stored in appropriate conditions (humidity, temperature, etc.)

MRI: Magnetic Resonance Imaging, application of analytical chemistry's nuclear magnetic resonance technique, used to produce real-time images, especially for examinations of the brain, spinal cord, heart, etc.

Wheatstone bridge: Wheatstone bridge, a circuit made of 4 resistive branches, used to detect a very small change in one of the resistors. Used to measure resistances with great precision, or tiny variations.

dpi: dots per inch, resolution of a printer representing the number of dots it can leave on one inch (corresponds to 2,54 cm)

Neural network: neural network, an integrated circuit that functions in a way similar to natural neural networks. (see theme by Rodolphe Bouichou)

TTL: Transistor-Transistor Logic, a logic signal usually between 0 and 5V

Common-mode rejection ratio (CMRR): for a differential amplifier, the ratio is the differential gain to common mode gain (often expressed logarithmically).

- **Internal architecture of the PIC 16F877:**

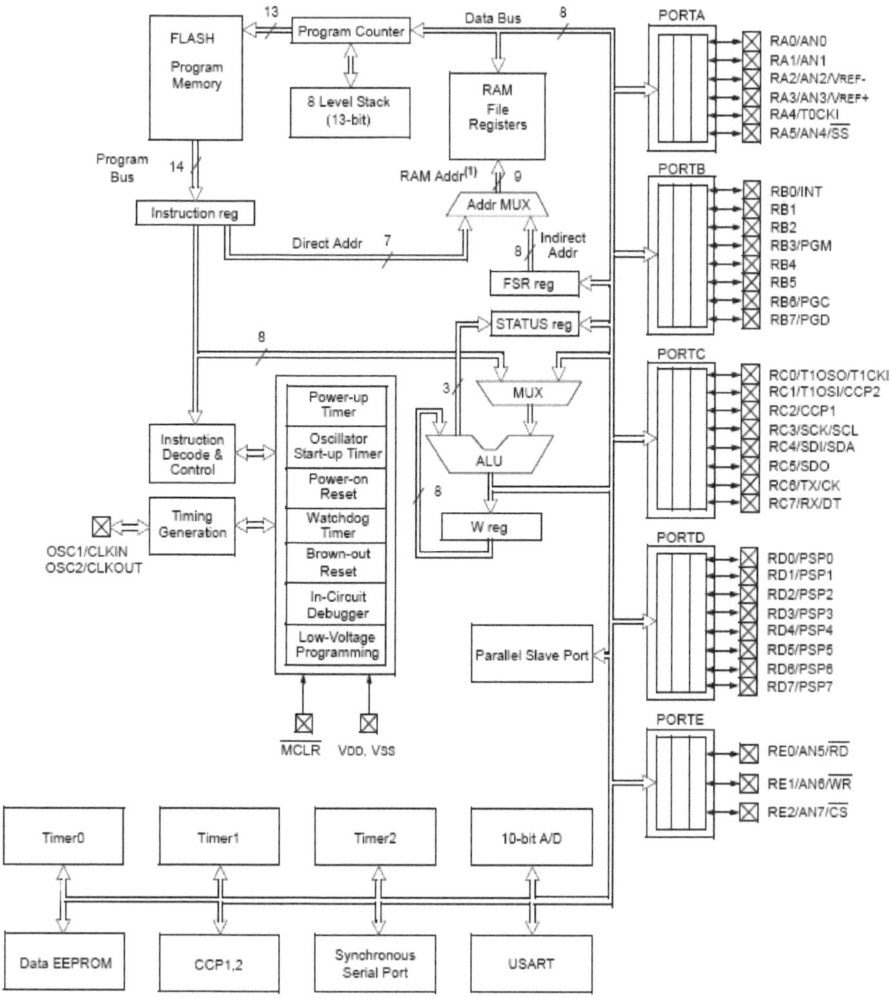

- . **The Arithmetic and Logic Unit:** program execution diagram.

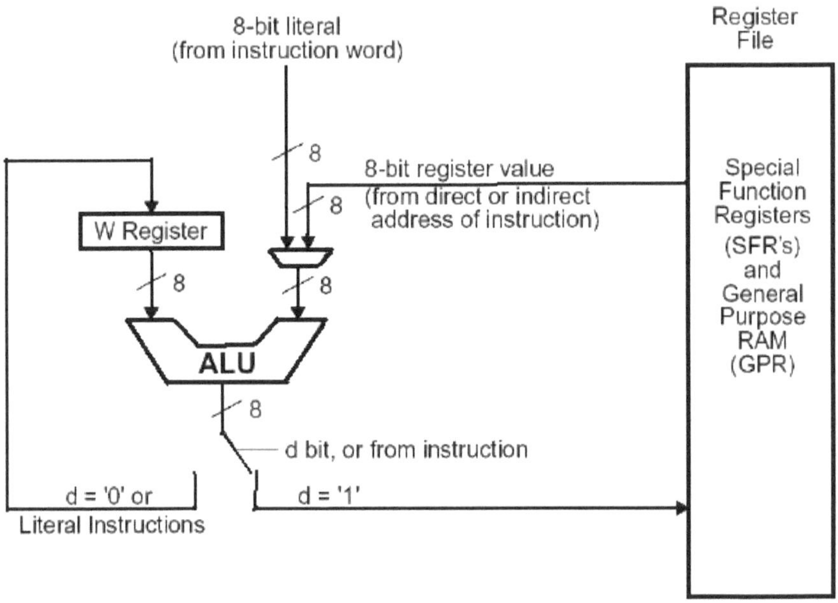

Kolibri 360 GmbH
Ferdinand-Jühlke-Straße 7
99095 Erfurt, Deutschland
produktsicherheit@kolibri360.de

Druck:
CPI Druckdienstleistungen GmbH
im Auftrag der
Kolibri 360 GmbH
Ferdinand-Jühlke-Str. 7
99095 Erfurt